The
Songs of Trees

Also by David George Haskell

The Forest Unseen: A Year's Watch in Nature

The
Songs of Trees

Stories from Nature's Great Connectors

David George Haskell

VIKING

VIKING

An imprint of Penguin Random House LLC

375 Hudson Street

New York, New York 10014

penguin.com

ISBN 9780525427520 (hardcover)

ISBN 9780698176508 (e-book)

Printed in the United States of America

1   3   5   7   9   10   8   6   4   2

Set in Minion Pro

Designed by Cassandra Garruzzo

Dedicated to my parents,
Jean and George Haskell

Contents

Preface

For the Homeric Greeks, *kleos*, fame, was made of song. Vibrations in air contained the measure and memory of a person's life. To listen was therefore to learn what endures.

I turned my ear to trees, seeking ecological *kleos*. I found no heroes, no individuals around whom history pivots. Instead, living memories of trees, manifest in their songs, tell of life's community, a net of relations. We humans belong within this conversation, as blood kin and incarnate members.

To listen is therefore to hear our voices and those of our family.

Each chapter of this book attends to the song of a particular tree: the physicality of sound, the stories that brought sound into being, and our own bodily, emotional, and intellectual responses. Much of this song dwells under the acoustic surface.

To listen is therefore to touch a stethoscope to the skin of a landscape, to hear what stirs below.

I sought trees in places whose natures seemed quite different. The first chapters of the book comprise stories of trees that seem to live apart from humans. Yet these trees' lives and ours, past and future, are twined. Some of these connections are as ancient as life itself; others are industrial reimaginings of older themes. I then turn to the exhumed remains of trees that have been long dead: fossils and charcoal. These ancients show the arc of biological and geologic stories and attest, perhaps, to the future. The third group of chapters concerns trees that live in cities and fields. Humans appear to dominate; nature seems absent or in abeyance. Yet wild biological relationships permeate every being.

In all these places, tree songs emerge from relationship. Although tree trunks seemingly stand as detached individuals, their lives subvert this atomistic view. We're all—trees, humans, insects, birds, bacteria—pluralities. Life is embodied network. These living networks are not places of omnibenevolent Oneness. Instead, they are where ecological and evolutionary tensions between cooperation and conflict are negotiated and resolved. These struggles often result not in the evolution of stronger, more disconnected selves but in the dissolution of the self into relationship.

Because life is network, there is no "nature" or "environment," separate and apart from humans. We are part of the community of life, composed of relationships with "others," so the human/nature duality that lives near the heart of many philosophies is, from a biological perspective, illusory. We are not, in the words of the folk hymn, wayfaring strangers traveling through this world. Nor are we the estranged creatures of Wordsworth's lyrical ballads, fallen out of Nature into a "stagnant pool" of artifice where we misshape "the beauteous forms of things." Our bodies and minds, our "Science and Art," are as natural and wild as they ever were.

We cannot step outside life's songs. This music made us; it is our nature.

Our ethic must therefore be one of belonging, an imperative

made all the more urgent by the many ways that human actions are fraying, rewiring, and severing biological networks worldwide. To listen to trees, nature's great connectors, is therefore to learn how to inhabit the relationships that give life its source, substance, and beauty.

Part 1

Ceibo

Near the Tiputini River, Ecuador
0°38'10.2" S, 76°08'39.5" W

Moss has taken flight, lifting itself on wings so thin that light barely notices as it passes through. The sun leaves not a color but a suggestion. Leaflets spread and the moss plants soar on long strands. A fibrous anchor tethers each flier to the swarm of fungi and algae that coats every tree branch. Unlike their crouched and bowed relatives in the rest of the world, these mosses live where water has no skin, no boundary. Here the air is water. Mosses grow like filamentous seaweeds in an open ocean.

The forest presses its mouth to every creature and exhales. We draw the breath: hot; odorous; almost mammalian, seeming to flow directly from the forest's blood to our lungs. Animate, intimate, suffocating. At noon the mosses are in flight, but we humans are supine, curled in the fecund belly of life's modern zenith. We're near the center of the Yasuní Biosphere Reserve in western Ecuador. Around us grows sixteen thousand square kilometers of Amazonian forest in a national park, an ethnic reserve, and a buffer zone, connected across the Colombian and Peruvian borders to more forest that, seen through the lofty gaze of satellites, forms one of the largest green spots on the face of the Earth.

Rain. Every few hours, rain, speaking a language unique to this forest. Amazonian rain differs not just in the volume of what it has to tell—three and a half meters dropped every year, six times gray London's count—but in its vocabulary and syntax. Invisible spores and plant chemicals mist the air above the forest canopy. These aerosols are the seeds onto which water vapor coalesces, then swells. Every teaspoon of air here has a thousand or more of these particles, a haze ten times less dense than air away from the Amazon. Wherever people aggregate in significant numbers, we loose to the sky billions of particles from engines and chimneys. Like birds in a dust bath, the vigorous flapping of our industrial lives raises a fog. Each fleck of pollution, dusty mote of soil, or spore from a woodland is a potential raindrop. The Amazon forest is vast, and over much of its extent the air is mostly a product of the forest, not the activities of industrious birds. Winds sometimes bring pulses of dust from Africa or smog from a city, but mostly the Amazon speaks its own tongue. With fewer seeds and abundant water vapor, raindrops bloat to exceptional sizes. The rain falls in big syllables, phonemes unlike the clipped rain speech of most other landmasses.

We hear the rain not through silent falling water but in the many translations delivered by objects that the rain encounters. Like any language, especially one with so much to pour out and so many waiting interpreters, the sky's linguistic foundations are expressed in an exuberance of form: downpours turn tin roofs into sheets of screaming vibration; rain smatters onto the wings of hundreds of bats, each drop shattering, then falling into the river below the bats' skimming flight; heavy-misted clouds sag into treetops and dampen leaves without a drop falling, their touch producing the sound of an inked brush on a page.

The leaves of plants speak the rain's language with the most eloquence. Plant diversity here reaches levels unrivaled anywhere on Earth. Over six hundred species of tree live in one hectare, more than in all of North America. If we survey an adjacent hectare, we add yet more species to the list. Every time I have visited, my anchor in this botanical confusion and delight is a *Ceiba pentandra* tree, a species that many local people call ceibo, pronounced SAY-bo. Twenty-nine paces take me around its base, steps that circle buttress roots radiating from the center, each root starting head high at the trunk, then sloping down into the forest. The trunk is three meters across, wider by half than the columns that support the Parthenon. Despite its impressive size, the tree is not nearly so ancient as the pines, olives, and redwoods that live in cold or dry climates and count their years in millennia. In the fungus- and insect-filled Amazon, few ceibo live more than a couple of hundred years. Ecologists estimate that this tree is between 150 and 250 years old. The tree is large not because it is old but because young ceibo lance upward by two meters every year, sacrificing wood strength and chemical defenses for speed of growth. The ceibo's crown, its uppermost branches, form a wide dome that rises ten meters higher than the surrounding trees, themselves forty meters high, the equivalent of about ten stories in a human building. From a perch in the crown I see a forest canopy unlike that of relatively smooth-topped temperate forests. A dozen other ceibos grow between my eye and the horizon, each one a hummock protruding from the uneven, fissured surface created by the surrounding trees.

The tree is a giant. An *axis mundi*? Perhaps, but the rain's sound refutes any attempt to use a single idea to isolate the tree from its community. Every falling water drop is a tap against leafy drum-skins. Botanical diversity is sonified, calling out under the drummer's beat. Every species has its rain sound, revealing the varied physicality of leaves of the ceibo tree and the many other species that live on and around its massive form.

The expansive leaflets of flying moss tick under the impact of a drop. An arum leaf, an elongate heart as long as my arm, gives a *took took* with undertones that linger as the surface dissipates its energy. The stiff dinner-plate leaves of a neighboring plant receive the rain with a tight snap, a spatter of metallic sparks. A rosette of lance-shaped leaves sprouts from the tip of a *Clavija* shrub, each leaf twitching as the rain smacks the surface. The sound is flat, *tup*, with none of the urgency of less yielding leaves. The leaf of an Amazonian avocado plant sounds a low, clean, woody thump.

These sounds come from the ceibo's understory plants, species that root themselves under the spreading branches and amid the duffy soil around the trunk. The water that strikes the understory has already passed across many leaves above. In the treetops most leaves have forms characteristic of the tropics: smooth surfaces ending in sharp tips or filaments. These "drip tips," combined with slick leaf surfaces, gather water, drawing it into large teardrops. As tears swell on the leaf tip, the water becomes a lens, refracting light so that an inverted image of the forest appears within. The drop has only a thin tip to hold, so every few seconds the leaf releases the accumulated water, then another lens bulges, flashing its image before falling away. The leaf thus sheds water, drying itself and slowing the growth of moisture-loving fungi and algae. These drip tips in the upper levels of the forest enlarge the already-giant raindrops, sending them down to understory plant skins. Larger leaves gather the most water and drip fastest, so the rhythms of the understory are born in the diversity of leaf shapes in the ceibo's crown. The myriad sizes, shapes, thicknesses, textures, and pliancies of the leaves below add texture to the sound. Even the litter sings with a vigor that I have not encountered elsewhere. This ground sound is the clack and tick of thousands of spring-wound clocks, each releasing its tension with a *tschak* unique to the woody muddle of the decomposing surface.

In the ceibo's crown, botanical acoustic diversity is present, but it is more subtle. Drops are smaller and create a sound like river rapids in the leaves of the many surrounding trees, obscuring variations in the sounds of individual leaves. Because I'm standing high up in the branches of an emergent tree, a tree that arches over all others, the sound of the river rapids comes from beneath my feet. I feel inverted, like an image in a teardrop, disoriented by hearing forest rain under my soles. My ascent, up a forty-meter series of metal ladders, has carried me through the rain layers: The sounds of rain on litter and understory plants fade a meter or two above the ground, replaced by the spare, irregular spat of drops on sparse leaves, stems reaching up to the light, and roots drilling down. At twenty meters up, the foliage thickens and the rapids begin. As I climb higher, the sounds of individual trees push forward, then recede, first a speed-typist's clatter from a strangler fig, then rasping drops glancing across hirsute vine leaves. I top the rapids' surface and the roar moves below me, unveiling patters on fleshy orchid leaves, greasy impacts on bromeliads, and low clacks on the elephant ears of *Philodendron*. Every tree surface is crowded with greenery; hundreds of plant species inhabit the ceibo's crown.

Human contrivances to keep away water are useless here and dull the ears. Rain jackets may repel falling drops, but their plastic magnifies the tropical heat and sweat soaks from within. Unlike many other forests, there is so much acoustic information revealed by the rain here that the crack, puff, or smack of drops on woven polyester, nylon, and cotton become an aural barrier and distraction. The yielding, lightly textured surface of human hair and skin is silent, or nearly so. My hands, shoulders, and face answer the rain with feeling, not sound.

When Western missionaries arrived here, they insisted that their colonized, evangelized subjects wear clothing. An unintended effect of this stricture was to reorient ears toward the self and away from the forest, partly closing the door to acoustic relationship with plants and animals. My conversations with members of the Waorani, the local indigenous culture, have almost without exception included unsolicited comments about the awkwardness and constraint of the clothes required when visiting town. The Waorani have lived in the forest for thousands of years, but outsiders now threaten their lives and culture. Clothes therefore weigh heavily for many reasons. I suspect that one of these is the disconnection from the acoustic community, a significant loss for people who live within multispecies relationships. Just as factory millworkers are deafened by machinery, given "cloth ears" by their work, wearers of the cloth itself also sometimes lose the ability to hear.

In the ceibo's crown, animal sounds overlay the plant rhythms: whine, murmur, howl, yelp, whistle, squeal, and whir. Every sonic verb has its champion, and many species communicate with sounds for which our language has no adequate descriptors. The blurred wings of a fork-tailed woodnymph hummingbird drone, a sound edged in whiplike squeaks. The bird, a thumb-size glint of iridescent blue and green, dips its beak into the red arch of flowers emerging from a zebra bromeliad. Between the bromeliad's fleshy pineapple-top leaves a frog calls *ko-ko-ko-UP!*, a jaunty song that awakens a call-and-response chorus from dozens of other frogs hidden in the thickets of bromeliads that cover the ceibo's branches. Unlike drip-tipped leaves, the erect rosettes of bromeliads collect and hold water. Each bromeliad can contain four liters in the gaps between the base of its leaves, a breeding site for frogs and hundreds of other species. One hectare of forest carries fifty thousand liters of water in treetop bromeliads, much of this volume pooled along branches of the large, emergent trees. The ceibo is a sky lake.

Water pools are not the only habitat in the crown. There are as many microclimates among these branches as exist within hundreds of hectares of most temperate forests. Bogs accumulate in sheltered crotches. Ephemeral wetlands fill and dry within knotholes. Dozens of years of leaf fall have accumulated in the ceibo crown, building soil as deep and rich as the litter on the ground. The soil lies on wide tree branches and is caught in tangles of vines. Rooted in this duff, a fig tree with a trunk as wide as a human torso grows amid half a dozen other trees in the confluence of ceibo branches, a forest rooted fifty meters above the ground. These trees cluster on the northern and eastern sides, where the crown's soil stays moist and the ceibo leaves grow thickest, like a shady forest ravine. On the exposed southwestern branches, a community of cacti, lichens, and razor-leaved bromeliads endure the alternation between deluge and desert, swelling in the rain, then crisping in the unobstructed equatorial sun. On vertical trunks, vines interweave with orchid gardens, creating water-holding mats in which ferns take root. Above all this grow the ceibo's own leaves, each one the size of a child's hand, fanning in eight or so elongate leaflets. The tree holds its leaves on twig tips, creating a gauzy haze. The leaves seem insubstantial for so large a tree, but unlike the sheltered plants below, these leaves must withstand the wind of thunderstorms and downbursts. Their small size and fanlike shape allow each leaf to fold up and yield to the wind.

Most tropical biologists have worked at ground level. But lately towers, rope ladders, and cranes have carried some scientists into the treetops. There they've found that as many as half, and perhaps many more, of the forests' species dwell in tree crowns and no-where else. "Canopy," the biological term for the crowns of many tree species in a forest, is too simple a word for such a complex, three-dimensional world.

Maps of biological diversity give us another way to understand the many lives of this ceibo tree. When the richness of plants, amphibians, reptiles, and mammals—admittedly a mere subset of life's diversity, but the subset that we know best—is tallied across the world, a color-coded map of diversity reveals the places with the largest number of species in each group. The place where these maps converge, the glowing eye of the map, is eastern Ecuador and northern Peru, the western Amazon. Tabular ranking of categories of species within these larger taxonomic groupings confirms what the maps suggest. By most measures this is the modern apogee of terrestrial biodiversity, the result of life's creativity incubated in tropical heat and rain. Evolution has had time to elaborate its hothouse productions: the western Amazon has been a tropical forest for millions, perhaps tens of millions of years. The region's geologic history is poorly understood, but the western Amazon's location between the uplifting Andes and the shifting coastline of the Atlantic may have opened it to invasion by new species from sea and mountains, further leavening the rise of biodiversity.

A less formal but equally informative indicator of the diversity of this forest is a walk with a professional botanist, either a professor or an experienced forest guide. Their extraordinary biological and cultural knowledge encompasses all the more common plants, including the role of plants in the lives of humans. The experts also bring specialized understanding of the identities and stories of a particular subgroup, the plants that they have studied for decades. But the task of identifying the majority of species, let alone knowing the stories of these plants, is far beyond their grasp. Species unknown to and undescribed by Western science are all around. Botanists recently discovered one new species on the walkway to the dining hall at a biological research station. This forest is the place where biological hubris dies: we live in profound ignorance of the lives of our cousins.

In the ceibo's upper branches, the rain eases. *Ara! Ara!* A pair of scarlet macaws pass directly overhead as they arrow from horizon to horizon, their flight an exultation of sound and color. At the tree singing insects divide the octaves, each species taking a place in the scale with clicks, wheezes, and pulsating buzzes. A plumbeous pigeon repeats a simple, low melody, joined by the titter and sneeze of a carnival of other birds: flame-crested tanagers, white-fronted nunbirds, blue-crowned trogons, at least forty avian species in a few branches. Howler monkeys call a kilometer away, sounding like a distant jet engine. Nine or ten other species of primate live here and punctuate the constant sound of singing insects with crash, whistle, and whoop.

Clouds turn to vertical wisps of mist, then disappear. The sun pushes down and the temperature climbs by ten degrees. Within two minutes my skin is dry; clothes will take days to turn from sodden to merely damp. A thousand bees descend on my body. Many of the sweat-drinking bees are small enough to pass through the mesh of the head net that I pulled on as the sun emerged. They plunge into my eyes with flailing, lacerating legs. After an hour or so of eye burn, I retreat from the bees' treetop realm, climbing down to the bipedal gloom.

As if descending back into Plato's cave, I'm changed in my return to the familiar world. Above there are biological layers of incomparable beauty and complexity. I'm in flatland now, but echoes and shadows of the superior layers play in my memory and on the forest floor over which I walk.

The sounds of the western Amazon never cease. The strands that connect life are wound so tight and packed so densely that the air thrums with vibratory energy night and day. Amid this intensity, the nature of life's network reveals itself in extreme ways.

At first this nature seems to be one of vigorous, even frightening, conflict. War cries and lamentations ring loud. The rule for humans in the ceibo tree or walking muddy trails: if you slip or need to steady yourself, do not grab the nearest branch. Bark here is an armory of spikes, needles, and graters. If, by luck, your grasp happens onto a

smooth-barked stem, the waiting ants and snakes will deliver the
lesson. Your laceration will fester in this aerial soup of bacterial and
fungal spores.

One does not need to reach out to find danger. I lean forward to
pick up my notebook and a bullet ant drops from the vegetation into
the gap between my shirt collar and nape, landing with a quiet *puk*.
Those curious entomologists who have deliberately sampled the full
palette of insect-induced pain rank the bullet ant at the top of their
global scale. The ant greeted my neck with a jab from a venomous
abdominal stinger. The pain was like a strike on a bell cast from the
purest bronze: clear, metallic, single-toned. I never knew how my
nerves could ring until the moment I was "lifted and struck" by
small-arms fire from a tree. My left hand slammed at the sting,
sweeping away the attacker. Before it dropped to the ground, the ant
scalpeled my index finger with its mandibles, slicing two grooves as
it bit down. Unlike the stinger's purity, this pain was a shriek, a fire,
a confusion. Over minutes the sensation ran across the skin of my
hand, a cacophony and panic that soaked the hand in sweat. For the
next hour my arm was incapacitated, my left pectoral muscle felt
wrenched and bruised. Hours later, muffled by drugs, the bite and
sting were reduced to a hot whine, as loud as a hornet sting but not
deafening. This was my initiation into one reality of the forest. I felt
none of Thoreau's "indescribable innocence and beneficence" in this
network of relationships. The art and science of biological warfare
reach their highest states of development in the rain forest.

The ant's assault left only a small scar on my finger. Other insects
leave more lingering and dangerous mementos. One of the quieter of
the insects that swarmed me in the ceibo crown was a murmuring
mosquito, its body a glinting royal blue, as large as a brooch. When
my attention was distracted, it sank its needle into my hand for a sup.
The lost blood was a trifle, but as it fed, this *Haemagogus* mosquito
drooled saliva into my capillaries, providing a liquid entryway for

viruses. *Haemagogus* is a treetop specialist, laying its eggs in damp crevices where rain awakens and sustains the larvae. The adult female's fondness for monkey blood, combined with her long life, makes the insect an excellent vector of disease. I was sharing a dirty needle with woolly monkeys. Or perhaps howler, saki, spider, capuchin, tamarin, owl, titi, marmoset, or squirrel monkeys? For a virus the treetops are swamplands of primate blood, joined by mosquito rivulets. Dozens of species of bat and rodent add tributaries. This mosquito species is a fecund home for viruses, bacteria, protists, and other blood-dwelling pathogens.

Luckily, no sylvatic yellow fever or other disease overtook me after my bite, but the mosquito was a reminder that although Tennyson's tooth and claw—pumas, snakes, and piranhas—grab our attention, most of the biological struggle in the forest happens at a scale that evades our senses. Samples of DNA reveal parasites in the blood and flesh of every creature. Only occasionally do we see an outward manifestation of this parasitism. As I listened to the dripping of water from a bromeliad, I saw an ant with its mandibles sunk into the outer edge of the leaf. The ant was dead. Its last act was the anchoring bite. A parasitic fungus, *Ophiocordyceps*, had consumed the ant from within, then somehow commanded the ant to crawl to a windswept leaf and grip tight. A stalk topped with a swollen sac now emerged from the ant's neck, spilling infectious fungal spores onto all ants below.

The leafy drumskins whose varied shapes turn the rain into sound also suffer multiple lines of attack. Bacteria and fungi bore through cuticles and breathing pores; insects gnaw on tender new shoots. In one of the better-studied plants, the *Inga* genus, half of the weight of the young leaves is composed of poison, a costly defensive investment. This is not a quirk of one botanical oddity; *Inga* is one of the more common and species-rich genera in the forest. Even with their poison, tender young leaves can sustain extensive damage, emerging from the vulnerable stage of growth looking like used shotgun targets. Older, tougher leaves have slightly less poison, but even they have up to a third of their weight invested in chemical defense, a reflection of the ubiquity of pathogens and the incessant nips and tears that herbivores impose on plants.

The severity of the struggle for life in the rain forest is both a result and a cause of the diversity of species. With so many species crammed together, competition is necessarily high and opportunities for exploitation abound. These antagonistic relationships feed the creativity of evolution, making the forest yet more diverse. If any one species becomes abundant, enemies swell their ranks and mow it back. There is an advantage in being rare: you give your attackers the slip. This rarity can be biochemical. If a plant is surrounded by close relatives but possesses its own blend of defensive chemicals, it will thrive despite living among plants that are similar in all other regards. Tropical plant communities are therefore extraordinarily diverse in part because the forest overflows with fungi and caterpillars. One hectare may contain sixty thousand species of insect, a billion individuals, half of which do nothing but eat plants and breed. Fungal and bacterial diversity and abundance are uncounted but likewise vast.

All this conflict would seem to force life into an atomistic mode. Individuals must fight it out, victim versus enemy, in endless, looping strands of conflict. The struggle is indeed intense, but instead of separating life into atoms, the Darwinian war has created a furnace that burns away the individual, melting barriers and welding networks as strong as they are diverse.

The culture of human societies in the region reveals some of this network. The Waorani have lived in the western Amazon for thousands of years, for much of this time as hunters, gatherers, and gardeners. Missionaries and other colonists brought disease and "assimilation," killing both people and culture. Now about two thousand Waorani live in and around the Yasuní Biosphere Reserve, some in permanent settlements with government schools and clinics, others in the forest, voluntarily isolated from contact with all other people. Waorani life within the forest has produced no Linnaean taxonomy of plants. Instead, many plant "species" have multiple names. Plants are often described by their many ecological relationships or uses within human culture, rather than by individual monikers. Anthropologist Laura Rival writes that when pressed by interviewers, Waorani "could not bring themselves" to give individual names for what Westerners call "tree species" without describing ecological context such as the composition of the surrounding vegetation.

Waorani society has no equivalent of the Himalayan cave-dwelling hermit or the Thoreauvian cabin, "living by the labor of my hands only." Waorani, in their own words, "live like one." Individuality, autonomy, and mastery are highly valued, but these are expressed in the context of relationship and community. Any who take to the woods to live in self-reliance are considered profoundly ill or angry, destined for death. Waorani "individual" names are a product of the group. Leaving one group for another entails the death of the old name, the acquisition of new personhood, and the impossibility of return.

To become lost in the forest, especially to be lost alone and at night, is a fearful event for Waorani, even those with the deepest experience of the forest. When Waorani do become lost, they find a ceibo tree and turn it into a subwoofer. Pounding on the buttress roots of the tree vibrates the whole trunk, a botanical basso profundo call to friends and family, a cry to reknit the bonds that keep you alive. The tree's great height lets it bellow in a way that shouting could never achieve. Hearing the pulsing air, your people will come. This signal is particularly helpful for lost children. Their families know where the large ceibo grow, so the sound both alerts and guides. Hunters and warriors also use the tree to signal news of kills. It is perhaps no coincidence that the ceibo is the tree of life in the Waorani creation story. The tree is a hub for so many forest creatures, and it saves lives by maintaining and reconnecting life-giving threads.

This dissolution of individuality into relationship is how the ceibo and all its community survive the rigors of the forest. Where the art of war is so supremely well developed, survival paradoxically involves surrender, giving up the self in a union with allies. Some of these alliances are forged within species. The bullet ant that shot me, the army ants whose legions cause the litter below the ceibo to tremble, the leaf-cutters that carry away bales of greenery to their underground nests—these are all societies whose identity resides in the colony, not the individual ant. My ascent of the ceibo carries me past many such arrangements. A tangle of spiderwebbing at the tree's base is the home of social spiders, a community of dozens, each spider contributing to the enlargement and defense of the web. Social spiders succeed or fail as a group. The characteristics of individual spiders matter insofar as they contribute to the community. Natural selection acts on these groups, favoring some mixes of spiders over others. Spider society therefore evolves through the fate of collectives. In a similar vein, many species of bird and monkey live in family groups united by mutual dependence.

Alliances that fuse distantly related species are just as common as those that form within species. The ceibo's roots and leaves are communities of mutualistic fungi and bacteria where the interests and identities of constituent parts become blurred. In the old, nutrient-poor soils of the Amazon such relationships are essential. Phosphorous is in especially short supply, and the ramifying network of the fungus's strands vastly increases the surface area available for absorption. The tree reciprocates with sugars from its leaves, allowing the plant/fungus union to thrive even in poor soils.

Fungi also sustain many ants. Fungus may have killed the ant on the bromeliad, but other fungi have hitched their fates to ant societies in mutually supportive ways. Leaf-cutter ants work on behalf of fungi, or perhaps the fungi are working for the ants. Union eliminates these distinctions. Columns of ants that stretch for dozens or hundreds of meters supply fresh leaves to fungus gardens in subterranean chambers. The ants feed the fungus and the fungus body feeds the ants. *Pseudonocardia*, a genus of bacteria that live among the ants' body hairs, keeps the fungus healthy by oozing chemicals that suppress interloper fungal species. The ant/fungus/bacteria convergence of lives has produced an entity whose essence is relationship. Any single part of this entity falls out of existence without interplay with the "others." Leaf-cutters are just one of more than two hundred species of attine ants, all of which depend on the cultivation of fungi for their livelihoods.

The hundreds of species of bacteria, protist, sponge, crustacean, and worm that live inside bromeliads depend on frogs that travel from one water pool to another. Ostracods, tiny shrimplike creatures, hold on to the frogs' skin. Clinging to these ostracods are ciliates, single-celled protists that feed on the bacterial broth within the bromeliad. At a yet smaller scale, bacteria and fungi ride the ciliates. All these creatures, along with the larvae of flying insects, defecate in the bromeliads' water, adding nitrogen and other plant-nourishing chemicals. Bromeliads therefore create and house their own dung farms. Like the leaf-cutter mutualism, bromeliad/animal/bacterial networks are twined such that most strands are inseparable. The forest is not a collection of entities joined by such networks; it is a place entirely made from strands of relationship.

Human culture expresses this nature in its philosophies. For people who have lived within the Amazonian forest's network for hundreds and sometimes thousands of years—the Waorani, the Shuar, the Quichua, and others—the forest is not an assemblage of biological and physical "others." Although these cultures are linguistically and historically divergent, and their belief systems are as varied as those over any continent, Amazonian peoples appear to agree on one thing: what Western science calls a forested ecosystem composed of objects is instead a place where spirits, dreams, and "waking" reality merge. The forest, including its human inhabitants, is thus unified. But this is not a union of what were separate parts; we exist from the start in spiritual relationship. Spirits are not otherworldly ghosts from a distant heaven or hell but are the very nature of the forest, earthed and grounded, connecting soil and imagination. Amazonian spiritualities grow from generations of pragmatic empiricism.

In thinking about these spirits, our English words and ideas fail us, coming as they do from another place. The barrier to understanding was expressed to me most clearly by Mayer Rodríguez, a forest guide who has worked with hundreds of American University researchers and undergraduate students. He said that not only would we not believe his stories of spirits but we *could* not understand. We can hear, but the sounds will not penetrate. The resonance of understanding is not possible without lived, embodied relationship within the forest community.

The relationships necessary for understanding extend back in time through genealogy and outward through space in webs of biological connection. Mr. Rodríguez's words give us a better outer understanding while conveying that comprehension from the inside will elude us. Knowledge is relationship; belonging is spiritual knowledge.

The Western mind can perceive and understand abstractions such as ideas, rules, processes, connections, and patterns. These are all invisible, yet we believe them to be as real as any object. Amazonian forest spirits are analogous, perhaps, to Western reality dreams such as money, time, and nation-states.

After one of my visits to the forest, I was part of a conversation with one of the Waorani men who first scaled the ceibo and built the tower of ladders that allowed me access to the crown. He is politically active and therefore unsafe, so he must remain unnamed here. While the tower was being built, he came to the tree at night, encircled the trunk with naranjilla fruit to hold the tree's jaguar spirit at bay, then talked with the tree, asking forgiveness. He built small fires to protect himself and the tree. As he related his story, he spoke of the tree as a person, not an object. The tree was violated when the bolts were drilled into its upper branches. A better tower, he said, would float among the tree's branches with no drilling or metalwork. It would carry Waorani children to the canopy and give them a place for music, for visual art. Years after the act, I saw only resigned sadness in the climber's eyes, but during the tower's construction he spent many days and nights in tumult. His coworkers, a mix of non-Waorani Ecuadorans and North Americans, were excited to construct an elegant access tower in a beautiful place. They'd done this before and chafed at his qualms.

Waorani are not opposed to harming or taking life. They cut plants, hunt monkeys and other animals, and defend their culture from foreign colonists and other Amazonian peoples with lethal efficacy. In settlements Waorani dependence on the forest is lessened by imported foods and more extensive agriculture, but here too the machete and the gun get much use. For the climber, harm to the ceibo was problematic for reasons other than a generalized opposition to cutting and killing. Ceibo is the life tree; "without it, we die," he said. Bolting the tree was harming and insulting life's source. I sensed that he also thought that the penetration of the Western mind into the canopy, made so easy with steps and guardrails, was dangerous for more subtle reasons. For visitors the tower is a means to an end, an expression of a particular philosophy of how to relate to the forest and therefore a statement about the essential nature of the forest. Building, then climbing the tower is an act weighted with moral significance; each clang on the ladder's rungs is the sound of a way of thinking, one that is often discordant with the philosophy of those who know the forest best.

Paradoxically, the tower allows outsiders to more fully comprehend the consequences of these different philosophies. From the tower's upper rungs we can hear and see the lands of the Waorani and the Quichua but also observe the expression of imported philosophies, ones that portend breakage of forest spirits on a scale far larger than the ladders on which we stand.

A tinamou sings the forest's vespers. Although this turkey-size relative of the emu is seldom seen, its melodies accent every dusk. The sound is the work of a silversmith, pure tones that the artist melts and crafts into ornament. The inflections and timbre of the Andean quena flute are surely inspired by the songs of these birds. In the understory the dark is comprehensive, but here in the ceibo crown, dusk lingers another thirty minutes, the orange gray western light of sunset reaching us unobstructed as we hear the tinamou's song.

As the light drains, bromeliad frogs spasm chuckles and grunts from aerial ponds. They call for five or more minutes, then cut to silence. Any sound will set them off again: a stray frog call, a human voice, the bleat of a roosting bird trodden by a companion. Three species of owl join the frogs. Crested owls punch regular groans from below, keeping in touch with mates, neighbors, and the youngster that the pair have hidden in the low branches of an *Inga* tree. The spectacled owl's repeated low, rubbery calls wobble around their crooked axis like a badly aligned tire. A distant tawny screech owl sings a high *to-to-to-to*, an endless, jabbing ellipsis. Insects pulse high drills, clear, sweeping chirps, saws, and tinkles. Monkeys and parrots whose sounds dominate the day have dozed away. The upper leaves of the ceibo chuff in the sharp gusts that accompany the sunset, then the wind eases and stillness comes to the tree.

It is now two hours after sunset. We're so deep in forest that the sky should be a dome of blackness and bright white dust. The people gathered here are a day's travel by road and river from Coca, the nearest town. Other than flashlights, there is no electricity except for the brief evening run of the dining-hall generator. Yet the sky is smeared by light from two horizons. Gas flares and diesel-fueled electric lights from oil-drilling camps just over five kilometers away are like the glow of towns, spilling into the black and dimming the stars. When the stir of leaves in the ceibo quiets, rumbles from generators and compressors wash through the tree's crown. The living riches of the western Amazon sit atop the graveyard of a Cretaceous shoreline. The sunshine that caused algae to flourish in river deltas and shallow seas 100 million years ago has left a buried oily residue. Maps that highlight eastern Ecuador and northern Peru as global high points of biological and cultural diversity align with maps of oil reserves. Oil worth tens, perhaps hundreds, of billions of dollars lies under these forests.

Half of Ecuador's export revenues and one third of the government's budget come from oil. After defaulting on bonds held by the West, the Ecuadoran government is now indebted to China, debts earmarked for payment with oil. For people in a country where many feel acute material need and want of economic opportunity, the sale of Amazonian oil seems a promising bridge to better lives, especially when loans and oil sales are channeled to social services. For the government, tapping the reserves is a ready source of cash and therefore reelection.

In most countries there would be little or no debate about drilling. In North America only a few oil fields stir any national controversy. Most are opened and exploited as a matter of course. The

North Sea is well tapped by the countries of northern Europe. Only wars and market calculations ease the flow from the Middle East. Yet in Ecuador, a country that by standard economic measures can ill afford to leave its oil unused, drilling in the Amazon has created a foment of protest and creative deliberation at all levels, from the president's office through civil society to small communities living in the forest.

In Ecuador oil belongs first to the government. There are no privately held mineral rights, although both private and state-owned companies receive concessions to exploit the oil. The government decides who drills where. The most controversial of these decisions concerns the area a few hundred meters from the ceibo tree, the forests of Yasuní National Park. The national park is one portion of the larger Yasuní Biosphere Reserve and covers nearly ten thousand square kilometers within the "hot spot" identified by maps of biological diversity. The park sits adjacent to the six-thousand-square-kilometer Waorani Ethnic Reserve. Some Waorani live within the park and ethnic reserve in voluntary isolation from other cultures. The northern half of Yasuní park is mapped into oil concessions. One of these, the Ishpingo-Tambococha-Tiputini (ITT) block, which includes the northeastern border of the park, contains an estimated 800 million barrels of oil, 20 percent of Ecuador's oil reserves. Yasuní also lies adjacent to oil fields that have already been exploited. In the 1970s U.S. companies turned large swaths of forest into oil-saturated industrial dumps. The courts are still wrangling over who is responsible for the long-delayed cleanup.

Access roads slicing through the forest have profound effects on the community of life. Hunters with new roads to market clear the forest of edible animals. Colonists take land from indigenous groups and turn forest into field and plantation. Where colonists are kept away by oil-company guards, some formerly mobile indigenous groups build permanent villages along the road. Many communities are fractured by debate over whether or not to work with the oil company. Disputes about who should receive company goods likewise cause dissent. Subsidies and employment can yield material benefits, but entry into the industrial economy often proves a short-lived boon as indigenous communities are displaced by colonists. Trees along Vía Auca, the main traffic artery leading to the older oil fields, have lost nearly every bromeliad, along with the animals that live within them. Birds that were formerly abundant shun the oil roads. Ceibo, if it survives the chain saw, loses its community and therefore falls silent. A Waorani man told me that oil drilling was like cutting the limbs from a ceibo, amputating the tree of life. Other Waorani have negotiated deals with the industrialists, trying to find ways to work with the latest swarm of outsiders coming to their land.

A few years ago it seemed that Ecuador might find a way to protect the forest, despite the stores of much-needed mineral wealth. In 2007 President Rafael Correa proposed that if the international community could fund sustainable economic development for just half of the value of the oil in ITT, Ecuador would leave the fuel in the ground in perpetuity. He later also proposed broader frameworks within the United Nations and OPEC for helping developing countries to manage fossil-fuel reserves and climate change. At the same time, the Ecuadoran government set new standards for its own actions. The 2008 Ecuadoran constitution protects the rights of Pacha Mama, the nature "of which we are part," including the right of nonhuman life to sustain itself and evolve, and the right of humans to live with access to water and healthy food. The proposal for Yasuní seemed a promising expression of this commitment.

Correa's plan would have kept oil drilling out of Yasuní and sealed unburned carbon in its tomb. This last benefit is particularly significant in a global context. If we are to have any hope of limiting the increase in global average temperature to two degrees Celsius, the stated goal of current climate negotiations, we must leave buried fuel in the ground. So even when we have the treasure map, we must turn away from the X. We have a lot of turning to do. Known global reserves of fossil fuels are three times higher than the amount that we could burn and stay within the temperature goal.

Correa's proposal failed. If Ecuador is to leave oil unburned, then Ecuador alone must shoulder the cost of lost opportunity. Those who have so far put most of the fossil carbon into the atmosphere, the citizens of deep-pocketed industrialized nations, were not willing to take on part of the financial burden of restraint. Buying oil, though, is easily accomplished. And so the ceibo tree daily hears machinery and is lit at night by columnar flames of waste gas taller than the grandest rain-forest tree. Seismic surveys are coming, ground-penetrating acoustic shocks in whose echoes the oil reveals itself.

Like any astute strategist, Correa had a backup even as he proposed the Yasuní plan. This alternative is now in action: develop the oil fields. In March of 2016 Petroamazonas, a state-owned oil company, sank the first oil wells in ITT, just north of the Yasuní park boundary. Among the political leaders of the region, Correa is not alone in his thinking. Over the western Amazon more than 700,000 square kilometers of forest have been mapped by governments into oil and gas "blocks." These areas comprise the majority of the Ecuadoran and Peruvian Amazon and large portions of the rain forest in Colombia and Brazil. The oil and gas in 60 percent of these blocks is now being extracted or explored, mostly along roads cut into previously roadless forest. A small number of extraction sites have no roads and are served by pipelines only, with access by air or boat. The remaining 40 percent of the blocks are in the promotional phase, as yet unleased to any oil company.

From the maps, future extensive oil extraction seems inevitable across most of the western Amazon. Ecuadorans have other ideas, though. A large majority favor keeping the drills out of Yasuní. A petition garnered over three quarters of a million signatures, well more than were needed to start a national ballot initiative. But the Electoral Council, politicized by Correa, declared most of the signatures invalid. Opponents of drilling are now hounded; defense of Pacha Mama will lose you your job, and more; the law courts are purged of the wrong flavor of judges. Such are the worries of many with whom I spoke. Dissent, they say, is now criminalized in the name of development.

Resistance to drilling takes many forms. Protesters march on Quito, nonprofits and academics release studies and press releases, activist outrage crackles through the Internet, and foreigners opine about how Ecuador should manage its affairs. What distinguishes this struggle from so many others is what lies at its heart: communities of people who are participants and listeners within the ecology of the world's most diverse forest. From these communities comes a philosophy of living whose words have rooted themselves in political discourse and the country's constitution. Thoughts from the forest, of the forest, have penetrated the nation-state.

Listening to these thoughts presents similar challenges as hearing and understanding the spirits of the forest. Our preconceptions erect muffling and distorting barriers of prejudice. In Coca, the oil town on the edge of the forest, racism does not hide itself. *Auca*—"primitive savage"—names a taxi company, Cooperativa de Taxis Auca Libre, a hotel, Hotel El Auca, and the main road to the oil fields, Vía Auca. At restaurants Waorani companions are treated with open disdain. To the south, Shuar and Ashuar alike complain of racial epithets. The Sarayaku Quichua are bullied by the military and attacked by anonymous thugs. Other prejudice comes wrapped in feelings of goodwill. Westerners seeking the "timeless wisdom" of forest peoples project their idealizations onto indigenous groups, not acknowledging that all cultures change, that all cultures are modern, whether they are rooted in the Amazon or in Athens. The revolutions brought by pre-Spanish intercultural warfare, the wholesale uprooting of peoples by the Incas, the decimation caused by old-world disease, the arrival of the Spaniards, hundreds of years of colonial machinations: all this happened before the industrial revolution ever stepped into the Americas. Since then the pace of external change has quickened. These outside factors combine with the interior evolution intrinsic to any culture. The myth of primeval humanity, untainted by modernity, is another way of slurring people with *auca*, failing to hear that every culture expresses its own form of modern identity.

Indigenous people in the Amazon, like all other people, draw on their history to understand the world, but such understanding evolves and is expressed to the outside in selective, pragmatic ways, colored by context and personality. The rain's sound is shaped and translated by drip tips, so what falls below is not the sound of rain but the sound of interpreted rain. All these preconceptions and misunderstandings challenge our ears, but they do not block all sound. As I spoke with people, I heard trees, or I think I did, through my distorting filters.

Teresa Shiki, a Shuar woman, a healer, activist, and teacher, ran away from missionaries who were feeding her bad food, teaching saints and statues, and forbidding her own language; she disappeared to the forest to find her grandmother. There she learned to listen to plants, to hear what they offered to humans. *Every tree is a living person, with speech. Ceibo represents all plant life; you cannot listen to "one" tree; there is no one tree living alone.* She listens as she walks; she listens as the plants speak in her dreams. *Our dreams are attached to the roots of plants, big and small, and to our ancestors. Oil extraction? The manifestation of a crazy mind, a mind living in lazy fantasy.* The industrial economy that she sees transforming her community is like a man running on hot coals, an unproductive, failing movement of escape. *This running takes you nowhere. When these terrible dreams come, turn to the ceibo, listen and live from within the tree, hold it close. Only with the tree's energy can we replenish our own spirits and have hope of survival. Only in wordless relationship to the tree can we receive this energy.* She's replanted a forest on degraded land and, with the Omaere Foundation, which she leads, shares the knowledge and medicine of the forest with locals and visitors, rebuilding the relationships from which people and the forest are made.

A Quichua man introduces his grandfather, son of a powerful shaman. The old man talks while his great-grandson winds strings of flashing lights through the needles of a plastic Christmas tree, a rustling fir. *The missionaries taught us the Bible, to write. We were no longer interested in trees. Before, we listened to the forest to hunt, to find animals. Now mostly we forget.*

The grandson relearns what two generations lost, the forest's words. He shares these with others, with visitors from across the world. *Alone among the trees, ceibo can withstand a storm; it gathers the wind in its wide branches and sends the violence down. This strength is what we lose when the ceibo is cut. Shamans are weaker now; many are liars. In the forest, away from the oil drilling and industry, the ceibo assembles and protects. Jaguar stores its food in the branches. Snakes and turtles lay their eggs in the soft soil below. Tapirs snout the same soil for the stink of rotted fruit. Snails, millipedes, bats gather on the trunk and in the recesses of the root buttresses.* He takes us to the largest ceibo left standing near town, ringed by cattle fields and homesteads. Snails and millipedes abound. The girl who lives next door says she hears spirits chattering like birds at night in the ceibo trunk and crown. They frighten her. *Sometimes God strikes a ceibo to kill the spirits within.* Missionaries, oilmen, and God: they work together.

In the center of town, Quichua men in suits work with and within the local government. *The central, national government hurts and kills the ceibo mother tree, cutting her away piece by piece. Even conservation programs encourage people to cut away the trees. We lose our medicines and hunting. State-driven conservation erodes the indigenous community. Without intact territory, owned and managed by the indigenous community, the forest falls into incoherence, the community dies. We go to the ceibo, hug the tree, and ask for strength, especially before dealing with the people from the oil and chemical industry. The sounds of the forest orient you, help you. They can make you happy or sad. Ceibo, like any tree, has her own sound; the touch and song of a large ceibo give us good energy.*

Another Quichua, a man who lives half in the forest and half in the national political struggle against industrial ruination of his people's land. *Trees have music in them. Rivers are alive and sing. We learn our own songs from them. People think we're crazy for saying that trees sing. It is not us who are crazy but those people who belittle us. Our politics is this: to show that trees and rivers have music, songs, and life; to turn the so-called national parks into living forests; to delineate our lands with gardens, filled with flowering, musical trees. This is not empty land; we've known tree songs for a long time, living with the millions of beings in the forest.* But the national "Law of Empty Lands and Colonization" says there are no people here.

Like mosses, the forest's thoughts can grow wings and fly. In Sarayaku, a Quichua community invaded by colonists and oil explorers, Carlos Viteri Gualinga and his colleagues craft words on paper to fly from the forest. As a defense against the many assaults on their community, they translate and politicize part of what they understand, publishing it in academic journals and political tracts. They reject the idea of linear progress from underdeveloped to developed, measured by the accumulation of material wealth. Instead, good and harmonious life—*súmac káusai, alli káusai*—should be "the goal or mission of every human effort." Such a life emerges from ongoing "reciprocity and solidarity" within the human community and between the human community and the biodiversity and spirits of the forest of which people are a part. Western development destroys these relationships, imposing itself by "blood and fire."

Their tethers to Amazonian trees cut, the Sarayaku papers alight in Andean Quito. With their home of origin unacknowledged, the words find a place in the nation's constitution, seemingly with only their spelling modified: "We . . . build a new form of peaceful coexistence in diversity and harmony with nature, to achieve *buen vivir*, the good life, *sumak kawsay*."

In the Andean air, in halls of government, *súmac káusai* loses the relationships in which it was born. Deracinated, it finds itself working for thoughts from elsewhere: socialism, sustainability, industrial economy. Amazonian *súmac káusai* becomes the nation's *buen vivir*, good living. Development is *buen vivir*; oil drilling will take us to a nationwide *sumak kawsay*. Forest thoughts flew to mountains, to the political heart of the country. They left as music from people, ceibo, rivers, soil. They wing back to the forest as the throb of drilling rigs and the fusillade of tires on gravel roads.

The rules of survival in the forest—reciprocity and solidarity—are tested. Now it is survival *of* the forest that is at stake. In the forest, the larger the assault and the more intense the warfare, the deeper the collaboration must be in order to thrive. And so human cultures whose relationships were formerly tense, even murderous, form cooperative networks. Frictions persist—cultural autonomy is strong—but the Confederation of Indigenous Nationalities of Ecuador is robust enough to transform the tone and content of the country's political discourse. Connections now stretch beyond national boundaries. Indigenous park guards talk across borders. Judges from across Central and South America, assembled at the Inter-American Court of Human Rights, hear the Sarayakus' case against the government and the oil companies. The judges rule in favor of the Sarayaku. The Ecuadoran government accepts and accommodates some of this power but pushes back against much of it. The vigor, even violence, of the reaction from the state reflects the strength of the alliances.

The art and science of warfare indeed reach a crescendo in the Amazon. If these were the only songs, the forest would spiral into annihilation. This is clear enough on Vía Auca. But the *súmac káusai* of life's community also emerges. The tensions inherent in conflict do not slacken, but their energies become so creative that mosses, frogs, and even the forest's thoughts take to the air.

Balsam Fir

I stand on a stony bluff overlooking a valley filled with the textures and hues of northern forests: blue green shades of fir needles, wind-stoked flashes of brightness from trembling aspen and white birch leaves, spiky crowns of spruce trees, somber canopy gaps over the stunted trees of bogs, and thickets of young evergreens where wind has leveled older stands. I'm on a trail at the edge of one such thicket, a growth so dense that no person could pass without a severe exfoliation. A balsam fir tree overtops most of the crowd of young trees, eight meters tall and about thirty years old. The fir's whole trunk is visible from the trail and its location on an elevated bluff yields breezes that, in the summertime, gave me intermittent relief from the hundreds of mosquitoes that gathered at my mammalian blood buffet.

A sound like fine metalwork rings from the top of the balsam fir. *Tink tink. Zreep.* Rivets tapped and rough edges filed. Birds rummage in the cones that swaddle the tree's apex. Their hammers never cease, unifying the flock, telling where seed is most abundant. As they work, shavings fall through the firs' branches, cone scales barely heavier than air, ticking against fir needles as they fall.

In summer the slate blue cones were clenched shut. Copious dribbles of resin kept away birds and squirrels. Now, in October, the cones have browned and the dried resin has fallen. Scales have eased apart to reveal stacks of thin, translucent paper. A flick of wind shatters the cone with a gentle snap and hiss, then paper kites stream away, some carried high, others spinning to the ground. Each kite has a traveler clinging to its base, a balsam fir seed barely thicker than the paper that carries it. Although the seeds are tiny, they are dense with energy. Drawn by these stores of food, birds join the wind, sweeping their beaks through the cones. The sunlight sequestered inside each cone is thus divided into hundreds of parts. A mossy bank receives the energy latent in a fir embryo, a pine siskin fattens its flanks, and nuthatches tap winter stores under bark crevices.

Of the birds that work the balsam fir, none are as vocal as the black-capped chickadees. The forest here is dense with fir, spruce, and pine. I can see no farther than a meter or two. But the chickadees' palaver advertises their location from tens of meters away. Like the restless movement of their bodies, chickadee sounds swing and hop, flickering through pitch and rhythm. They punch the air with guttural *deer deer*, then ascend an octave and give a quivering two-noted squeak like the vigorous rubbing of glass. High jabs intersperse slurs, then the voice drops to a throaty *chik-a dew dew*, the call that we humans used to give the species its name.

In every season that I visit the balsam fir tree, the chickadees flock me. Whether they are assessing, greeting, or just passing through by chance I do not know. Their inspection is thorough. One arrives and jumps its calls to the higher registers, then half a dozen more birds gather around me. I freeze. They perch on bouncing fir twigs, centimeters from my face, tilting and ducking their heads as they pass their impenetrably dark eyes over me. Their voices rasp as they wing from one side of my face to the other. I see them as they must see one another, not as distant shapes in a treetop but as beings of great visual intricacy: a tracery of gray plumes over their shoulders, blade-edged flight feathers, combed felt on their cheeks. Sometimes other birds are drawn to the gathering, perhaps responding to a change in the character of the chickadees' acoustic news ticker. A northern parula warbler comes, then a magnolia warbler and a redbreasted nuthatch. These others glance, then drop out of sight. The chickadees are more curious and linger for minutes, then return to gleaning insects from fir needles or poking at cones. These are brief visits, unexceptional for them, I expect, but these chickadees are bolder and more inquisitive by far than any others I have encountered. Most striking are the fine variations of timbre and inflection that emerge from my close hearing of their chatter. At this intimate distance, a seemingly single type of call—*deer*—resolves into many sonic variants.

From twenty-six simple geometric shapes we've constructed a written language; in a few minutes of attention to their flock, I hear perhaps as many acoustic graphemes. We have only a weak grasp of how these sounds texture the birds' experience of the world. Some calls predominate during breeding or are given near the nest. Other sounds transmit information about danger, using slight acoustic variations to encode information about the threat posed by different predators. Nuthatches eavesdrop on these variants, gleaning knowledge from their chickadee neighbors about which species of predatory owl are present in the forest. The chickadees use many other sounds as the birds interact with one another, seeming to convey the subtleties of affinities and disputes. No doubt our language and their communications diverge in many ways, but heard closely, the two are not so different in their acoustic richness.

My inspectors are a social species. Their intelligence resides within both individuals and societal relationships. A chickadee therefore lives in a dual world, a self and a network. These birds are just one example of the larger duality within the forest's nature, one that permeates the biological world and may date back to the origins of life. Chickadee lives echo the stranger world of the balsam fir tree, the forest, and the creative ambiguities of biological networks.

Inside their skulls, the sophistication of the neural capacity of black-capped chickadees increases in autumn. The part of the brain that stores spatial information gets larger and more complex, allowing the birds to remember the locations of the seeds and insects that they cache under bark and in clusters of lichen. The superior memory of the birds that I hear in the tips of the fir tree is a neuronal preparation for the hungry days of late autumn and winter. The seat of spatial memory in the brains of chickadees that live in these northern forests is particularly voluminous and densely wired. Natural selection has worked winter into the birds' heads, molding the brains so that chickadees can survive even when food is scarce.

Chickadee memories also live within societal relationships. The birds are keen observers of their flockmates. If one bird should happen on a novel way of finding or processing food, others will learn from what they see. Once acquired, this memory no longer depends on the life of any individual. The memory passes through the generations, living in the social network. If black-capped chickadees are like their European relatives, regional traditions color this cultural knowledge. Birds in one part of the forest may favor a particular style of opening cones or capturing insects, a style transmitted to them from the happenstance learning of their ancestors. Generations ago, a bird in the western part of the forest may have discovered a faster way of extracting a fir seed. The bird's eastern counterpart also invented a new cone-breaking technique, slightly different from the western approach. Now the innovators are dead, but west and east still differ, even though the two methods are equally effective. These traditions trump individuality. Birds conform to their group's preferred habits even when they have successfully tried the other way of feeding.

Bird behavior is of great importance to the balsam fir tree. Although the majority of the tree's seeds are wind dispersed, bird beaks are often the blow that sends cone scales flying. The birds' hunger has two opposing effects on the trees' future. The balsam fir's reproductive efforts are trimmed by the thorough work of foraging birds, a loss for the trees. In eating the seeds, birds divert energy that could have nourished young seedlings. The trees' stores are rerouted into bird stomachs and keep wild gray flames on the wing. This robbery is a heavy burden; it takes fir trees two years to build the energy needed to produce a full seed crop. But in hiding the stolen loot through the forest, chickadees and other birds lodge fir seeds in rotting logs and other prime seed beds. In winter many of these seeds are recovered and eaten, but some are forgotten. Bird memories are therefore a tree's dream of the future. The Amazon is not the only place where the unseen dream is real. Despite its neurological abstraction in the mind and culture of another species, chickadee memories are as important for the balsam fir as are soil, rain, and sunlight.

The chickadees' two ways of holding knowledge "in mind," in the individual and in the network, mirror the principles of the balsam fir tree's own intelligence and behavior. Even though it lacks a nervous system, the tree's cells are awash in hormones, proteins, and signaling molecules whose coordination allows the plant to sense and respond to its surroundings.

Some plant responses are long term, such as the growth of branches into light or roots into fertile soil. Plant architecture is not a haphazard affair but is the result of constant assessment and adjustment as conditions change. Twigs sense the luminosity of their particular location on the tree and grow accordingly. Flat fans of needles grow in the shade, to maximize exposure to meager sunlight, but in strong sunlight needles take on an upswept form to both gather sun and minimize shading of needles below. Branches offset themselves vertically from those around them, avoiding shading and wiggling themselves into sunlight.

Other responses last just a few minutes. The upper surface of a fir needle is a waxed floor, an unbroken green sheen. Underneath, two silver lines run lengthwise down the needle. Seen through a magnifying lens, the blur of silver resolves into a dozen rows. The rows are wheat-crop straight, hundreds of bright white dots on a green background. These dots are breathing pores, each one made from the gap between two curved cells. The cells integrate information about the state of the needle's internal environment, then open or close pores to admit gases or release water vapor. Every cell inside the needle is making similar assessments and decisions, sending and receiving signals, modulating its behavior as it learns about and responds to its environment.

When such processes run through animal nerves, we call them "behavior" and "thought." If we broaden our definition and let drop the arbitrary requirement of the possession of nerves, then the balsam fir tree is a behaving and thinking creature. Indeed, the proteins that we vertebrate animals use to create the electrical gradients that enliven our nerves are closely related to the proteins in plant cells that cause similar electrical excitation. The signals in galvanized plant cells are languid—they take a minute or more to travel the length of a leaf, twenty thousand times slower than nerve impulses in a human limb—but they perform a similar function as animal nerves, using pulses of electrical charge to communicate from one part of a plant to another. Plants have no brain to coordinate these signals, so plant thinking is diffuse, located in the connections among every cell.

The balsam fir tree also remembers. If caterpillars or moose browse its needles, the nibbling assault lodges itself in the chemical makeup of the tree, in a manner analogous to the changes in a chickadee's nerve cells after a near miss with a predator. The tree's subsequent growth is more heavily defended by unpalatable resins, like a bird turned jumpy by its bad experience with a hawk. The fir also remembers air temperatures dating back nearly a year, a memory that helps the tree to know when to winterize its cells. Plant memories can cross generations, as the offspring of stressed parents inherit an enhanced capacity to generate genetic diversity when they breed, even if this next generation experiences benign conditions. We only partly understand how plants hold these memories. It seems from experiments with cress plants that changes in the proteins that wrap DNA may be partly responsible. By looping DNA either tightly or loosely, plants can store information about which genes will be most useful in the future. Plant memory is thus captured in biochemical architecture.

Roots and twigs have memories of light, gravity, heat, and minerals. Darwin discovered some of these abilities by rotating young bean roots and showing that they remembered their previous orientation for many hours. He compared the roots' behavior to that of a headless animal, with memory suffused through the body. Whether the balsam fir has exactly the same abilities as the beans and cress plants is unknown, but the tree possesses the same internal chemical and cellular networks as these laboratory-grown species.

Part of a plant's intelligence exists not inside the body but in relationship with other species. Root tips, in particular, converse with species from across the community of life, especially with bacteria and fungi. These chemical exchanges locate decision making in the ecological community, not in any one species. Bacteria produce small molecules that serve as signals, allowing the cells to make collective decisions. These same molecules soak into root cells, where they combine with plant chemicals to promote growth and regulate the architecture of roots. Roots also signal to bacteria, providing them with sugars that both nourish the bacteria and switch on their genes. This halo of food and encouraging chemical signals causes the bacteria to cluster in gel-like layers around the root. Once established, the bacterial layer defends the root from attack, buffers it from changes in salt concentrations, and stimulates growth.

Roots converse with fungi, sending chemical messages through the soil. On receiving the message, symbiotic fungi grow toward the root and reply with their own chemical ooze. Root and fungus then change the surfaces of their cell membranes to allow more intimate contact. If the chemical signals and cell growth occur in the right sequence, root and fungus entangle and begin an exchange of sugars and minerals. In addition to food, the root chimera moves information from one plant to another in the form of chemical signals that travel through the fungus. These molecules carry messages about attacking insects and drying soil, the stressors of plant life. The soil is therefore like a street market. Roots gather to exchange food and in doing so they also hear the neighborhood news.

Nearly 90 percent of all plant species form belowground unions with fungi. Our eyes therefore tell half-truths when we gaze on a forest, a prairie, or a leafy urban park. The verdure of plants that we see is only part of the network that brings the community into being. For many trees, especially those like balsam fir that grow in cold, acidic soils, the fungus/root relationship is particularly well developed, comprising a sheath of fungal tissue around every root. Working together, both fungus and plant can thrive in the challenging physical environment of boreal forest soils.

This network of communication also includes leaves. There plant cells not only sniff the air to detect the health of neighbors but also use airborne odors to attract helpful caterpillar-eating insects. Sound plays a role in this communication. When a leaf senses the vibrations of a caterpillar's moving jaws, these chewing sounds cause the leaf to mount a chemical defense against the insect. Leaf cells therefore integrate chemical and acoustic cues as they sense and respond to their surroundings.

A leaf is not just composed of plant cells, though. The waxy surfaces of leaves are peppered with fungal cells, and leaf interiors are host to dozens of fungal species. Like fungi in the roots, leaf fungi have smaller cells than plants and lack photosynthetic pigments. Fungi are closer kin to animals than they are to plants, gaining their food not from sunlight but by absorbing food into their bodies. This suggests a reason for the ubiquity and success of intimate plant/fungus relationships. The two partners are sufficiently different that each can bring a talent that the other lacks. The union melds two different parts of life's family tree, yielding creatures whose physiology is nimble and many-skilled, both in leaves and in the soil. A leaf populated with fungi is able to deter herbivores, kill pathogenic fungi, and withstand temperature extremes much better than one composed only of plant cells. There may be one million species of leaf-dwelling ("endophytic") fungi on the planet, making them one of the world's most diverse groups of living creatures.

Virginia Woolf wrote that "real life" was the common life, not the "little separate lives which we live as individuals." Her sketch of this reality included trees and the sky, alongside human sisters and brothers. What we now know of the nature of trees affirms her idea, not as metaphor but as incarnate reality. Like the union of leaf-cutter ants, fungi, and bacteria below the ceibo, a tree's root/fungus/bacteria complex cannot be divided into little separate lives. In the forest, Woolf's common life is the only life.

Outside the laboratory, the complexity of the relationships among trees and other species increases by many orders. Decisions are made in these networks based on flows of information involving thousands of species. The chickadee's culture looks simple in comparison. It is therefore not just the balsam fir tree that thinks but the forest. The common life has a mind. To claim that forests "think" is not an anthropomorphism. A forest's thoughts emerge from a living network of relationships, not from a humanlike brain. These relationships are made from cells inside fir needles, bacteria clustered at root tips, insect antennae sniffing the air for plant chemicals, animals remembering their food caches, and fungi sensing their chemical milieu. The diverse nature of these relationships means that the tempo, texture, and mode of the forest's thoughts are quite different from our own. The forest, though, also includes humans, chickadees, and other nerved creatures. A forest's intelligence therefore emerges from many kinds of interlinked clusters of thought. Nerves and brains are one part, but only one, of the forest's mind.

As the sounds of foraging birds tumble from the highest part of the fir tree, clinks and jangles emerge from the ground. A ruffed grouse struts out of a thicket of balsam fir and spruce seedlings. The bird's steps are fox silent on the rotting needles, then crackle as its feet pass over the trail. My own footfall is like the grind and punch of walking on a sidewalk strewn with shattered glass. Even tree roots evoke sound. The swell of growing roots causes shards of rock to click, a sound so quiet and soil-muffled that I detected it only with a probe nestled into the rocky ground. The brush of a fingertip on the probe is a roar compared with the tick of rocks nudged by roots. Some botanists suggest that the quiet sounds made by roots stimulate plant growth, but these claims are controversial. Too few human ears have attended to the soil's chatter, and experimental evidence is ambiguous. So at present we don't know whether these sounds are inconsequential side effects of growth or meaningful conversations, analogous to the well-known chemical signals that travel among roots.

The crack-voiced earth around the fir tree is made of hard, brittle stone, alternating layers of black chert and rusty iron. A few of the layers are pencil-lead thin, but most are as thick as a human finger. Chert is a mineral made almost entirely from silica, so under my touch it feels like glass and fractures into slick-sided blocks with edges sharp enough to cut skin. In skillful hands these blocks can be worked into knives and scrapers. Such tools, marked by distinctive black and rust banding, are the only physical evidence left by the first Paleo-Indian colonists of this land. Subsequent Indian cultures made more sophisticated tools—adzes, points, and chisels—that relied on the chert's keen edge. Europeans later brought new uses. Run across steel, a confusion of sparks streams from the chert's slicing edge. Early rifles used these chert-on-steel "gunflints" to ignite a pinch of gunpowder, a "flash in the pan" whose sparks would pass through a narrow bore to light the main charge packed inside the weapon. The geologic formation that the fir tree's roots press against is named for the old firearms. The Gunflint Formation arcs from north-central Minnesota across western Ontario. The fir tree grows near the formation's center, thirty kilometers west of the city of Thunder Bay.

Between the chert, iron lies in laminar deposits that, when exposed to rain on this mountain slope, trickle away in dirty rivulets. In places where the rock is recently exposed—in landslides and along eroding trails—the land looks like a scrap yard of rusting stone, heaps leaching their iron. Downstream, rivers are the color of overbrewed tea, stained by iron from rocks and tannins that dribble from forest soils.

Nearly two billion years have passed since these rocks settled on an ocean floor. At the time, rising levels of oxygen were oxidizing iron that had floated freely in the ocean. This oxidized iron, rust, settled out of the water into thick beds. Worldwide the process created "banded iron" deposits that are now the geologic signature of this part of Earth's history. So much iron was sequestered into rocks that we now turn to these formations for ore. Several large iron mines are scattered through the Gunflint.

Examined closely, the chert layers between the rusty iron reveal the source of the oxygen and the cause of the iron deposition. Etched into the chert's fine silicate crystals are filaments and spheres, imprints of long-dead cells. Until marks in an older rock in Australia unthroned them, these were the earliest known fossiliferous signs of life. Most of the cells were photosynthetic, using sunlight to weld carbon into sugar, bubbling oxygen from their brazing torches. The fir tree's roots, the grouse's feet, and my boots were all sounding our deep biological history, ringing at an unmarked geologic memorial of Earth's first life.

Darwin knew nothing of such fossils. In his day the fossil record extended to the Cambrian, to about 600 million years ago. That no fossil antecedents existed to the large, complex animal forms of the Cambrian was an "inexplicable" puzzle to Darwin, one that he believed was a "valid argument" against his evolutionary ideas. It was not until the 1950s that the Gunflint fossils were discovered, a find that more than tripled the known age of life on Earth. Australian finds have now added more than a billion years to this span. Life is at least three and a half billion years old, truly as ancient as Darwin suspected.

The Gunflint fossils hide themselves well. The chert is as black as ebony wood, deeply stained with carbon. This carbon is a hint that remains of life may reside in the rock. The fossil cells are fifty times smaller than our unaided eyes can resolve. To get a clearer view, paleontologists direct the thrumming power of an electron microscope's beam at the rock, then pass the reflected energy through computer software that creates three-dimensional images. Thanks to the publication of images from this modern microscopy, anyone with an Internet connection can see the Earth's ancient cells with the same precision that Darwin and his contemporaries applied to animal bones and other human-scaled artifacts.

Compared with the forest that grows over their remains, the

fossils' community was composed of a small number of species, a couple of dozen at most. No multicellular creatures existed. Many cells were strikingly similar to the filaments of modern photosynthetic bacteria, others were simple globes, and a few were equipped with stringy arms or thick capsules. Despite this modesty of size, diversity, and form, the Gunflint community prefigured the most important relationships and life processes of the complex species that would evolve later. Like human music and art, life established its motifs and relationships early. Paleolithic flutes made from griffon-vulture bones—the first known manufactured instruments— were tuned to the same pentatonic scale used by many musical traditions in the post-vulture-bone eras. Painters from the same time, creators of the beasts leaping across Lascaux's walls, had, according to Picasso, "invented everything." Biology, like music and the visual arts, is a story of improvisation and elaboration on germinal themes. In the Gunflint this theme is a tension, one that has dominated life since: the creative tug between the individual and the community, the atom and the network.

A few of the Gunflint cells drifted in the water as plankton. They floated over a muddy bottom covered in slimy mats composed of a loose assemblage of several species. Life had already divided itself into a community in which varied species took on different ways of life, some seemingly quite individualistic, some more physically entangled. The most common of the fossil remains in the Gunflint chert are not just entangled but fused. These fossils need no microscope, at least for a first glance. Seen from above, the fossils lie in a mosaic whose tiles range in diameter from a few centimeters to more than a meter. Each tile lies atop another of the same size, so the visible mosaic is just the surface layer of columns packed against one another, descending as much as a meter. Each column is a stromatolite. When it was alive, the stromatolite was coated with a weave of microbes that grew like a city, building new layers on the sediment left by previous generations. Over hundreds of years, microbes lifted their homes from lowly villages to closely packed towers.

The stromatolites' living tissues rested on the upper surface, bathing in sunshine. Coils and strings of photosynthetic bacteria, *Gunflintia*, formed the majority of the living fabric. Larger globes of other bacterial species, *Huroniospora* and colonies of *Corymbococcus*, lived within the strands, like herbs emerging from a bed of moss. Burrowed within this green profusion were tiny spheres whose chemical compositions suggest that they were grazers or decomposers of other species. A dozen other cell types of obscure function dwelled in the Gunflint stromatolites. The fossil remains of these life-forms evince intimate ecological bonds and mutual dependence.

Warm lagoons in Mexico and Australia are home to modern living stromatolites and, although these modern forms have acquired new species, we can glimpse the dynamics of the Gunflint communities through these present-day analogs. Like a miniature ceibo crown, each fraction of a millimeter of the layered surface of a living stromatolite supports a different species. These diverse members of the community feed on the chemicals produced by their neighbors. Interdependent relationships among different species give the community its fundamental character. Chemical gradients and flows of electrons animate the living mat. At night the community switches from feeding on sunlight to processing sulfur, changing its internal chemistry to match. If the Gunflint stromatolites were anything like their modern counterparts, they were biological compositions in which the vitality of each note depended entirely on the phrase in which it was embedded. Two billion years ago, the boundary between the self and the community was already blurred.

Was an individual a single cell of *Gunflintia*, or a filamentous chain of such cells, or the whole stromatolitic disk? Perhaps such a search for individuals, for the "units" of biology, is misguided. The fundamental nature of life may be not atomistic but relational. The essence of the Gunflint community is the network of interactions, not the collection of selves. Any single answer to these questions denies part of the reality of these microbial microcosms. Life has now turned the whole planet into a stromatolite. A thin film of networked organisms spreads over the surface of the Earth rock, building on the rubble of past ages.

The chert-rooted fir tree is one filament in the planetary membrane. The tree seems an exemplar of individuality, its vertical trunk the antithesis of reticulation. The fir indeed grew from a single embryo within a single seed, its DNA coding a unique genetic identity. When the trunk falls, this individual will pass away, a biological atom with a beginning and an end. But like all trees, the fir's separation is also an illusion, seen only when we tilt our head one way. Every needle and root is a composite of plant, bacterial, and fungal cells, a weave that cannot be unknotted. The singular fir embryo was planted by a bird whose feathers were a sheen of bacteria, whose gut was a microbial community, and whose context was a cultured society. Cracked open, germinated, the seedling's growth was possible only because no herbivorous moose swallowed the young tree into a four-chambered mash of digestive microorganisms. For the absence of moose, the fir owes its life to wolves, human hunters, and the mosquitoes that infect moose with nematode worms and viruses. Modern stromatolitic green—the forest in which the fir tree grows— calls down its own rain, seeding the sky just as the Amazon does. The air's chemistry aggregates the aromas of pine, spruce, and fir into particles that draw mist into droplets. These add to the dust, smoke, and exhaust of North America's air, and a fine rain descends. The fir tree's life is relationship. Tilting our heads away from the atom, it seems that life is not just networked; it *is* network.

The tension between atom and network reaches further back in time than the Gunflint. The stromatolite community preserved within chert is old, but its cells already had a billion or more years of evolution behind them. Life's origin is buried yet deeper. For decades biologists have defined life as a *self*-replicating process. The search for a biochemical explanation of life's first steps has therefore been a search for stable molecules that can faithfully copy themselves. A few such molecules exist, most notably some RNAs, chemical relatives of DNA. These molecules are animate origami. In folding themselves they create new copies, form and function united in one molecule. If life started this way, it truly would be a triumph of individualism. But chemical networks offer an alternative model of life's origin. These networks are collections of relationships that, once established, regenerate the network, not any particular individual. The simplest example is a triad. Molecule *A*, instead of replicating itself, brings molecule *B* into being; *B* then begets molecule *C*, which then generates *A*. In the lab these networks can assemble themselves from rudimentary chemical precursors, then beat self-replicating molecules in a Darwinian contest.

The first artificial cells also have a networked character. When scientists organize chemical reactions into arrays of tiny, interconnected compartments, lifelike properties emerge: cycles of protein production, gradients of signaling chemicals, and the ability to maintain a steady internal state. Every cell in our bodies does the same. The geometry of the network in synthetic cells determines the speed of reactions, the rhythms of oscillations, and the manner in which signals are produced. Without the network, the homogeneous chemical soup lacks any tang of life.

The same lessons about networks are now emerging in the biotechnology industry. In the early days of DNA engineering, scientists manipulated single cells to accomplish relatively simple tasks. For example, they inserted a single gene for human insulin into one strain of bacterium. The descendants of this modified bacterium became drug manufactories, living in a carefully controlled broth of food, ceaselessly producing insulin. But for more complex tasks this focus on the individual has proved inadequate. Genetic engineers have been unable to develop single strains of cells that can turn wood into liquid biofuel or clean complex mixes of pollutants. Instead, networks of cooperating cells, each engineered to interact with the others, can accomplish what no individual can. Away from the laboratory, the world is yet more complex. All of life's ecology and evolution is animated by networked relationships.

Chemical relationships seldom fossilize—old chert truly is opaque—so we'll likely never know exactly how life started. But networks appear to be more evolutionarily robust and fruitful than selves. Networks fend off competitors, they enliven the chemistry of cells, and they persist through time.

A network, once established, might be called an individual. But this individual's character is defined by a collection of relationships, not by the stable presence of any particular molecular identity or genetic code. The particularities of relationships change over time—a feedback loop to D may be added and A become optional—but the network lasts and is the essence of the life-form. So life has a contradictory, creative duality in its nature: it is atom or network; it is neither and both. This is not a question of metaphor; it is life's fundamental character. Life straddles two states of existence and thereby animates what was a dead universe.

In passing from a chemical melee at the origin of life, through the middle ages of the Gunflint, into the modernity of a forest, life's network not only persisted and diversified but also acquired cells and bodies thousands of times larger than the wisps and specks of its early years. Most modern creatures—the microbes that dominate the biological world—never trod the path to multicellularity. They live, as did their ancestors, in an anarchic community of ever-changing coalitions and rebellions. In a small number of cases, though, a federal system took hold. Microbes aggregated, lifted themselves from the microbial swamp, and, still coated with their anarchist relatives, swam, crept, and walked over the stromatolites into the world of gigantism.

Seas rich in oxygen provided the physical conditions needed to sustain the hungry metabolism of these new creatures. Oxygen, though, did not solve the bigger problem of how to assemble life into even more stable, coordinated communities. In the fir tree the reproductive interests of the needles, the roots, the bark are entirely subsumed into a larger collection of cells. This is an unstable arrangement. Long-term persistence is threatened by conflicts between the interests of the larger group and the interests of smaller collections of cells within the group. The cancer of unrestrained cellular individuality can, and still does, destroy networks from within. In laboratory experiments, bacteria will spontaneously form multicellular aggregations that benefit all the cells involved, rudimentary versions of plant or animal bodies. Then some bacterial cells mutate and reap the rewards of community without reciprocating with the investments needed to hold the community together. These freeloaders thrive for a short time but then become so abundant that the whole aggregation falls apart. How, then, to tighten the stitching so that a fir tree or a chickadee is possible?

Very occasionally mutations confer a benefit to the group at a cost to the individual. These mutations are the stitches that cause biological aggregations to become more closely knit. Specialization—a cell taking on a single role in the body—is usually explained as an advantageous step toward efficiency. A cell that focuses all its efforts on being a good leaf, cone scale, or root might outperform a generalist cell. A less obvious benefit is that any mutation that increases the specialization of cells closes the door to cellular individuality. A solitary root cell cannot thrive. But a root cell connected to a leaf cell might be a successful Darwinian innovation, one that has no possibility of turning back to the selfhood of single cells. The genetically preprogrammed death of cells is another mutation that devastates the evolutionary prospects of a solitary cell but confers advantages on a group. Our nervous system would be a misfiring tangle were it not for the pruning brought about by the self-sacrificial death of extraneous cells. Our toes and fingers would remain fused without the death of embryonic cells between each digit. These cellular changes—specialization, programmed premature death—are unpickable embroiderers' knots. Once tied, life's strands cannot slip. The weave tightens.

At minus forty degrees Celsius, the cold becomes animate. No longer a mere sensation, it is a presence, a forceful consciousness muscling against my boundaries. Whether I sit or stand or walk in place at the fir tree, the cold's grip strengthens, the clench of a boreal wrestler. The hold sends streaks of burning chill across my face, down my back, through my hands. I limit my sedentary watches to a couple of hours each, then stride, run, or take shelter in town, loosening the wrestler's lock hold.

The cold not only bears down on human bodies but also bends sound. The forest sits under an inversion, chilled air pooling under a warmer cap. The colder air is like molasses for sound waves, slowing them as they pass, causing them to lag sound traveling in higher, warmer air. This difference in speed turns the temperature gradient into a sound lens. Waves curve down. Sound energy, instead of dissipating in a three-dimensional dome, is forced to spread in two dimensions, spilling across the ground, focusing its vigor on the surface. What would have been muffled, distant sounds leap closer, magnified by the jeweler's icy loupe.

The horn of a freight train blasts in my ear. The tracks are nearly an hour's snowy trudge away, but this morning the diesel engine and steel wheels run directly past my feet. Truck engines gunning up a rise on the Trans-Canada Highway, tire rubber spinning on ice, the aggressive whine of snowmobiles: they're all among the fir trees, mingling with the *churr* and *chip* of red squirrels and chickadees. Here are modern and ancient sunlight, manifest in the boreal soundscape. Squirrels nipping the buds of fir trees, chickadees poking for hidden seeds and insects, all powered by last summer's photosynthesis; diesel and gasoline, sunlight squeezed and fermented for tens or hundreds of millions of years, now finally freed in an exultant engine roar. Nuclear fusion pounds its energy on my eardrums, courtesy of life's irrepressible urge to turn sunlight into song.

The train was headed east, most likely hauling grain from western Canada to the town-size complex of silos in Thunder Bay, one of the world's largest grain ports. From here cargo ships haul the prairie's seed across Lake Superior into a global web of trade. In the Thunder Bay Museum, colored ribbons pinned to a world map show the connections: Asia, Europe, Africa, and the Americas. Global embroidery.

Alongside the grain silos, and piled nearly as high, are log stacks and wood pulp. These heaps feed wood-pellet factories, sawmills, and paper manufactories. In the cold air, steam from a large paper mill rises as high as the mountain ridges to the south of town, dominating the skyline, a painter's dream of ever-changing color, texture, and form. At close range the sounds are just as rich. A steely heartbeat of conveyor belts keeps rhythm, gas pipes wheeze and sigh, and the engines of Canadian Pacific locomotives drum their piston fingers as they wait to unload. Behind metal-plate walls, a gurgle and rumble. Trees are digested to pulp, then squeezed flat. As with all paper mills, I smell the maw and guts of the factory long before I hear them. The russet, hot scent of pulverized wood is stained with plumes of swampy hydrogen sulfide. Like grain hauled from the west, wood from the boreal forest flows through this port to many parts of the globe. Canada leads the world in sawn-wood production and is second to the United States in wood pulp. Worldwide about 10 percent of these two classes of wood product come from Canadian forests.

The edge of the boreal forest in which the fir tree stands is a place where humans have created particularly intense flows of matter. Today fuel, grain, and wood are the forms of embedded biological energy that move most vigorously. Two hundred years ago, fur and tobacco were the currencies of exchange. Trappers from across central and northern Canada gathered in summer to trade pelts for twisted braids of leaf. The fir tree stands on one of these old trade routes, a portage around the forty-meter drop of Kakabeka Falls. Workers would slog up the Gunflint rust–stained trail, each man carrying two forty-kilogram packs to the canoes on the Kaministiquia River, a water highway to the interior. Through their work, Virginian tobacco leaf found its way into a forest from which hundreds of thousands of pelts were hauled to Europe. Beaver was valued for the felted hats that were made from its undercoat, but skins of every furbearer in the north came through: muskrat, fox, otter, bear, wolverine, even seal from the arctic. The fur trade soon collapsed and the local economy reoriented to mine and timber exports. These are echoes of older human connections. In precolonial times Indian copper from the region made its way to South America and ceramic-making skills flowed northward. Pitch from balsam fir held and waterproofed the seams of the birch-bark canoes that made possible much of this exchange. Trade and knowledge were carried by aromatic resin.

People followed the furs, the ore, the wood, first through trade routes, then colonial landgrabs, and lately through long-distance immigration for industrial jobs. The global weave represented by colored ribbons of grain routes is also manifest in the cultural diversity of this trade hub. Fort William First Nation sits adjacent to the paper mill. The nation is an Ojibwe island, surrounded by the floodwaters of land taken and occupied by colonists. The camps and forts of the French and British colonists are gone, but they lie buried nearby, along the rivers that floated all trade. In the modern town I eat salt fish in a Finnish cafeteria, hearing Finn-English from the chatter of the retirees around me. Down the road is the Italian Cultural Centre and the church of St. Casimir, where services are offered in Polish. In summer the Festival of India brings bhakti dance to the waterfront, against a backdrop of cargo ships from Germany and Hong Kong. Elvis sings "Funny How Time Slips Away" at a diner decorated with Graceland kitsch. None of this is far from a rail line, a lumberyard, or a dock.

These connections and movements are extensions of the solar-powered network, mediated by the actions of *Homo sapiens*. We continue the patterns of the Gunflint community—flow, communication, interdependence, outgassing—and although the vigor and scale of our actions are reweaving the world's network with alacrity, such profound change is nothing new. Stromatolites brought about one revolution: the burn of oxygen from *Gunflintia* killed any microbe that happenstance had not shielded from the novel chemical. Then the stromatolites' descendants overpowered their own ancestors, smothering them, chewing away at layers of bacterial community. As a consequence, stromatolites are now confined to a few backwater lagoons where competition is minimal. The first trees to evolve also usurped many of their elders, grasping light before it could reach the trunkless plants below. The extra oxygen released by these ancient forests catalyzed the evolution of flying insects and other large animals. All these transformations were disruptive; all involved changed relationships. My frostbitten ears hear variations on old themes as diesel locomotives pull railcars of seed and trucks haul metal ore overland.

Biological networks are seldom quiet for long; usurpation and revolution come, creating and destroying. Thoughts, textures, and rhythms die. These are painful losses for those of us who love the melodies we were born into. In the unfamiliar, jarring sound that emerges, we hear discord and, possibly, a segue to new harmonies.

In 1972 the Landsat satellite, a truck-size marvel of New Astrology, was hurled into orbit. No longer would we peer up at shifting patterns of stars to augur the future. We had a star of our own. In 2013 the eighth Landsat satellite was launched, a continuation of the longest-running space-based study of the Earth's vegetation and terrain. These satellites skate the sky, circling the globe every one hundred minutes, recording the scene below with electronic sensors. Like a combine working a wheat field, the path of Landsat's orbit is arranged in offset swaths to cover the entire field, the globe. By projecting trends from the last decades, we see through a satellite's glass, darkly, and squint into the future.

The lidless eye sees both the flush of new growth and fields of stumps. Bare ground outstrips new growth. Summed over the globe, the area of land covered by forests is plunging. The first dozen years of the millennium saw 2.3 million square kilometers of forest lost but only 800,000 regrown. In the boreal region losses outstripped gains by more than two to one, the result of fire and logging. Government statistics obscure these patterns, tallying "forest" wherever young trees might grow, even if no trees are present. Landsat's pictures do not run through the filters of creative accounting. They report a boreal forest in retreat.

Landsat's images have a resolution of thirty meters; they are painted with a fat-tipped brush. But the forest community is a filigree drawn with the finest pen. To understand the satellite, we have to come back to ground. I've returned to the fir tree in summer and, except at night, when the cool air pools and the sound lens returns, the trains and trucks have left the forest. Instead, the wind directs a chorus of trees. Aspen leaves shudder when the air moves slowly, then spasm into pattering chaos in more forceful gusts. A little calmer, drier, are the white birch leaves that lift from tap to fizz as the wind picks up. These deciduous trees almost smother the rustling, chafing sounds of the fir. The balsam fir tree holds its stiff needles apart. These living bristles are silent, except in the highest winds. But browned, fallen needles caught in the living foliage rub against the thick, dangling shag of horsehair, antler, and rosette lichens that drapes every branchlet in the tree. These combs and tangles grate as twigs bob and the main trunk sways. Dead needles and cone scales drop, *tik*, into the moss below. Higher wind speeds invigorate the abrasion. The tree hisses like fine steel wool burnishing a tabletop, a sound that is strong, corrosive, but with a soft bite.

The fir's summer song is one of dead matter, moss, and lichen, seemingly minor parts of the forest network. Our human senses—and therefore our sense of what is important—are tuned to louder creatures, not to the murmurs of fallen needles and dreadlocks of moss or lichen. We deceive ourselves, though, if we don't occasionally turn from eagles, squirrels, and aspen to examine the duff and dross of forests. A study of these recondite members of the community unveils how changing forests connect to global cycles of energy and matter. Landsat's data finds its meaning in the soils and "lower" creatures of the boreal forest.

Soils in boreal forests hold three times as much carbon as all the forests' tree trunks, branches, lichens, and other aboveground life combined. Roots, microbes, and decaying organic matter are therefore a vast repository of carbon. Depending on the details of accounting methods, boreal soils are either the world's greatest terrestrial carbon store, outweighing even lush tropical forests, or they come a close second. Worldwide, soils contain three times as much carbon as the atmosphere, so the future of our climate depends on the fate of hissing and rasping fir needles. If the carbon encased in these falling needles should turn skyward instead of lodging in soil, our warming blanket of carbon dioxide would turn to a well-stuffed, overheating quilt.

The enormity of the boreal carbon reserve is partly due to the vastness of the forest itself. One third of the world's remaining forests grow in the boreal. But even if we ignore their extent, the forests are still disproportionately rich in carbon. When dead needles and mosses lie in cold, water-saturated soils, decomposition is sluggish and a backlog of dead matter soon accumulates. For much of the year, the ground is frozen, paralyzing the microbial activities that turn solid matter into airy carbon dioxide. When summer's meager and short-lived warmth returns, the microbes are again slowed, this time by sodden, acidic conditions. As I stand at the fir, the hundred iridescent-winged mosquitoes that hold me in a cloud of soft, humming wingbeats attest to these swampy conditions.

Conditions in winter and summer conspire to build soil carbon. Over the thousands of years that have passed since the last ice age, at least 500 petagrams (500 thousand million metric tons) of carbon have accumulated in boreal soils and peatlands. We get a glimpse of this carbon in the gardening aisle of a shopping center: the pallets of "peat moss" stacked to the ceiling are flecks of boreal and arctic carbon, taken from boggy soil, then shipped south.

The boreal forest is warming much faster than the rest of the globe. Fires have become more frequent, accounting for much of the recent loss of forest. In a fire, not only is carbon in the soil incinerated, but the remnants are left unprotected after flames destroy covering vegetation. As fires release carbon to the atmosphere, the boreal forest turns from a carbon "sink," a place that absorbed and stored carbon, into a carbon "source," a place where the net flow of carbon is from the soil to the atmosphere. Because atmospheric carbon is a greenhouse gas, the conversion of the boreal forest from a carbon sink to a carbon source adds yet more eiderdown to the atmospheric quilt.

Less visible than fires, but potentially just as important, are the changes in the soil's network of relationships. Warmth sends soil microbes into a frenzy. Their activity increases exponentially as soil temperature increases. If the warming persists for days or weeks, the composition of the community changes, further accelerating activity as cold-adapted microbes are replaced by heat lovers. The result of these changes is faster rot. Dead needles, roots, fungi, and microbes are processed through the soil's living community and their remains sent to the sky. The biological fire makes no smoke, but it is all-pervasive and therefore more important to the global flow of carbon than the drama of flame.

The supply of nitrogen also affects the vigor of decomposition. When nitrogen is limited, microbes slow their work and carbon builds up in the soil. This state of mild nitrogen starvation is the normal condition for microbes in most of the boreal forest. The mantle of lichen and moss that covers every surface in boreal forests intercepts and captures nitrogen from rain and dust, stopping nitrogen from reaching microbes in the soil. But when the lichen and moss communities are gone, after a fire or after herbicide treatments for forestry, nitrogen flows to the soil microbes unhindered and acts like a shot of caffeine for the process of decomposition.

The relationships among roots, fungi, and microbes also modulate the effects of nitrogen. In the boreal forest, most tree roots are fused with fungi that specialize in sucking nitrogen from the soil. The trees gain a source of carbon and the fungi are rewarded with sugary gifts from the trees. The microbes living in the soil away from roots, though, lose out. The root/fungus partnership grabs nitrogen that the soil microbes could have used to power their business of rotting the dead. Where such root/fungus partnerships thrive, microbes are listless and carbon builds up in the soil. The boreal is one such place. Farther south, tree roots associate with different kinds of fungi and they do not draw down the soil's nitrogen reserves. These southern trees are already moving into the boreal, pushing northward as temperatures rise. If this continues, yet more of the boreal forest's carbon will move from soil to sky.

Sitting on the moss and chert under the fir tree, I can sense the forest's behavior in many ways: through the tap of falling cone scales or the bellow of a train, in the society of roots or the cultural memory of chickadees, and through the remembered abstractions of carbon budgets or Landsat pictures. In all these I encounter the continuation and elaboration of the Gunflint's ancient network of living thought. How this thought moves into the future will depend on the relationship among needle, root, microbe, fungus, and human.

In the boreal there is reason to hope that we'll guide the human part of this relationship with forethought. Over the last two decades, continent-wide planning for conservation, forestry, and industry in the boreal forest have brought people together who have fought for years in the law courts. Now timber companies, industry, conservation groups, environmental activists, and governments, including those of the First Nations, are talking to one another. These conversations take many forms, manifesting in agreements, frameworks, initiatives, panels, and councils. Such human talk is part of the forest's larger system of thought, one way that the living network can achieve a measure of coherence: a diffuse conversation, able to listen and to adapt. To date, swaths of boreal forest as big as many countries—hundreds of thousands of square kilometers, more than 10 percent of Canada's boreal forest—have been mapped for conservation, for carbon-savvy logging, for threatened animals, and for sustainable timber production. In some places relationships among parties in the negotiations are strained; as in the rain forest, conflict is part of the network. But perhaps more important than the details of any map or agreement is the multiplication of connections among people. The boreal forest can now benefit from the diversity of experience within the human community and, through this, our varied ways of understanding the ecology of the world.

Below the fir tree, at the base of the cliff, chickadees rattle and hiss. An immature bald eagle screams. Awkward wings slap at treetops as the eagle lifts. Ravens spy the ungainly youngster and they rise in its wake, plunging, twisting, swirling around their quarry. The eagle's ponderous wingbeats are no match for the raven's agility, but the chasers hold back, seemingly content to play but not to strike. They follow until the eagle crests a ridge, then return to their roost on the hillside near the fir. *Kwok-kwok*, repeated dozens of times.

From the black chert have risen networks of chemistry, of biology, and now of culture. Intelligence stirs the air as the ravens parley. Memories twine as a seed connects chickadee to fir tree. My pen scratches on milled wood, another rootlet, minding the forest.

Sabal Palm

St. Catherines Island, Georgia
31°35'40.4" N, 81°09'02.2" W

Newtonian spheres trace rings through the void. Earth and Moon hoop the Sun, setting the earthly rhythms of day and night. Moon circles a spinning Earth, each tracing an arc in the other's sky. The spheres would fling themselves apart were it not for threads of gravity that interconnect all mass, whether that mass be a star, a moony dust mote, or a drop of ocean.

A bulge of water chases the Moon around the planet. The ground also feels the moonward gravitational pull, but rocks are too stiff boned to leap. The oceans are more responsive, raising tides in answer to the Moon's draw and the Earth's spin. At any ocean shoreline, the interlocking orbital rings manifest themselves in flood and ebb. All the fueled cunning of humanity cannot move this volume of water, a heave of weighty ocean, yet the rotating spheres emanate a silent power that emerges from nothing but relationship.

When, in their rotations, the spheres temporarily align along one axis, the combined gravitational pull of the Sun and Moon cause the oceans to rise high and fall low: the spring tides, leaping, springing. Days later Moon and Sun are out of alignment, at cross-purposes; gravitational forces are anemic, causing gentle neap tides.

Imagined from within the abstractions of celestial geometry, water's movement is orderly, imbued with mathematical elegance. Even when the overtones and ornaments of irregular shorelines and ocean depths are worked into the score, all seems harmonious: Earth and ocean are governed by the steady, predictable hands of the skies.

No sunlight, no Moon. A storm pounds offshore. I hear nothing but the violence of water. A few waves hiss; most give a deeper complaint as they charge, then punch. Embayments and spits impede and deflect the assault, causing waves to turn on one another, releasing slaps so loud that they resonate in my chest. Every few seconds, lightning cracks the dark: surf sliced by a giant oak that lies dead on the beach; spilling breakers overtopping beaten, limp palm crowns; sea spray so dense that the lightning fires the air with silver. Then darkness. At my feet, shudders emerge from what was steady ground. Waves slam into the knee-high escarpment that marks the highest edge of the beach; body-size fragments of soil cleave away; the roots that held the soil are entirely powerless. The moon presses the tide so tight against the land that spent waves have no room to run back before the next breaker arrives. By my clock, the tide is at its highest point; it should ease back soon, but my whole being tells me, *You're next.* There is no celestial harmony but atonal panic, sensory tumult that overwashes all else. Not Newtonian elegance but Prospero's rough magic and roaring war.

These spring tides, drawn up the beach by the full moon, felled the sabal palm that I have visited every few months for the last two and a half years. Tonight I discovered that the tree had fallen. Every wave soaks the upturned root-ball and ocean water drowns fronds that, a few days ago, stood atop a nine-meter-tall trunk, lush and vigorous. The fronds were talkative, full of rustle and snap. Now I hear in them only the detonations and bellow of the sea's quarrel with the land.

I'm on a beach facing the Atlantic, on St. Catherines, a barrier island off the U.S. coast in Georgia, with 6,500 kilometers of ocean between me and the western coast of Morocco. The island lies in the center of the wide arc of the southeastern coast, the Georgia Bight, a curve that stretches from North Carolina to Florida. Waters here are shallow, so as tides sweep in from the seaward northern edge, the advancing front of water swells as it gathers in the ever narrower, shallower concavity of land. On St. Catherines the ebb tide is three vertical meters lower than the flood tide. In Miami, south of the bight, the tidal range is less than a meter. The island therefore receives the magnified force of the tides and waves coming from the Atlantic. When a high tide combines with the surge from a winter nor'easter or a late-summer tropical storm, the modest chunks of land that I see fall away are much larger. One wave will take out a cliff or an entire dune.

The sabal palm is not the only creature killed by tonight's tide. Seawater surges through wrack and sand dunes, fanning into the land behind the beach's upper ridge. To get to the palm, I struggled through thickets of saw palmetto, relearning how these toothy plants were named. The suck of waves pushed and tugged my boots, even though this land normally stands twenty meters behind the beach. When the tide drops, what were freshwater lagoons, oak and palmetto forests, and hibiscus-filled meadows will be sand covered, their soils drenched in salt. One breach can open a channel that kills many hectares of wetland or smothers a large copse of forest. Ninety-nine percent of tides don't reach this high, but those other tides count for nothing if the 1 percent chew gashes and spit salt into the land's edges. Once a tide reaches this point, what were terrestrial communities will shortly be turned to beach, then seawater. Over the last century and a half, the land here has retreated, depending on location, by between two and eight meters each year.

It is not just spring tides and storms that push back the land. Two and a half years ago, when I first encountered the sabal, it was rooted several meters behind a dune. It stood in a row with half a dozen other sabal palms, all emerging from an untidy hedge of saw palmetto. Behind them palmetto thickets intermingled with groves of wind-stunted live oak trees, some with trunks over a meter thick. The seaward edge of the dune was sliced away, creating an escarpment as high as I am tall. The beach ran from the toe of this steep slope. The slack remnants of waves from the flood tide occasionally lapped close to the dune, but the palm was a dozen meters beyond the tide and, because the dune was lower on its beach side, ground level for the palm was more than a meter above the beach. The palm seemed secure behind and atop its ramparts. As I sat through calm summer days, listening and watching, I saw what an illusion this security was. Even on a windless day at ebb tide, the dune was receding through the accumulation of thousands of tiny losses.

At that time, the sand on the seaward edge of the dune banked at a sharp angle, running to the beach in one sweep from just below the dune's peak. Sitting close, I heard this face whisper, a sibilant hesitation, only audible when the seethe of distant wavelets quieted for a few moments. The sounds came from liquefied sand, patches of the slope that suddenly lost their grip and turned, in an instant, to fluid from granular solid. The sand hissed as it raced down the slope in narrow chutes. As the flows hit the beach, the sand huffed as it fanned. Some rills lasted just a few centimeters before friction arrested them. Most traveled the full length. Flows came about one per minute. The slope looked uniform and solid, but gravity spoke otherwise and unlocked first one cluster of grains, then another, dotting its force across the dune face with no discernible reason or pattern. A beetle struggling up the slope unleashed dozens of slippages and a dangling blade of dune grass incised an arc below which the sand was all fallen away. In one afternoon, along a two-meter stretch of beachfront, the North American continent lost a bucketful of land to the action of beetle feet, grass blades, and the fickle grip of sand grains. Summed across all dune faces on the coast, a bevy of dump trucks was at work, tipping sand into the sea.

It took one year for storms and beetle feet to remove the dune. The sabal palm now stood at the top of the beach, still firmly planted among its companions, with just a few of its easternmost roots exposed. Shallow, slack water from spent waves of the highest tides eased around these roots, but no violent water came. The tides left the sand smooth and bare. A sharp line marked the beach edge. Behind this, around the palm trunk, was a mess of leaf mulch, sand, and grass roots, soil with the look and smell of earth. Toads, deer, and lizards foraged in the litter but never set foot on the salty beach.

When waves approach the shore, they heighten as the sea bottom below them rises. The deepest part of the wave drags along the sand, but the upper wave feels no such restraining force and pushes on. The swell finally overtops itself in breakers that pound their energy into the sands. On a beach with a gentle incline, some of this energy launches a sheet of water toward land, a bubble-filled sluice that loses speed as it advances, finally halting, then sagging back into the sea. At their highest point, these laps of seawater are barely deep enough to wet a human foot. They are gentle, easing between my toes.

A hydrophone—a microphone protected within a waterproof, egg-shaped rubber shell—reveals a different experience for sand grains and palm roots. What my feet felt as a tender hum was, sensed from within the water, a clamor. I had expected mere sloshing but got my ears blasted when I sank the hydrophone. A lap of seawater arrived with the impact of a bucket of water hurled at a wall. Immediately I twisted down the volume on my sound recorder. The scrape of water across sand was like the sweep of a planer across wood, then the sound rose to a shriek as the sand grains accelerated. As the water receded, it entrained a growl of dragged, jostling grains. The touch of the sea, the most gentle of water movements, overpowers any sand that it reaches. Grains tumble and fly. Lighter particles, like clay or pieces of dead leaf, are swept away. Roots that previously held the soil are washed clean. The beach is flattened by the superior force of water.

What my unaided human senses felt in a stormy spring tide is what sand grains experience in even the calmest of weather. Like beetle feet and gravity acting on a dune's face, the graze of wavelets reshapes the coastline. Unlike the maw of a storm, these nibbles continue day and night, year-round.

Humans are a species adapted to landscapes that stay stable through our lifetimes. We are attracted to sturdiness and longevity in both the land and our homes. Building on rock is wise; fools choose sand for their foundations. The application of human inventions to the land—concrete, steel girders, plate glass—enforces the illusion of a changeless world. Instability unsettles us: fallen monuments, crumbling homes, and leveled forests are sites of pathos. Places that suggest permanence or durability—the thousand-year-old stone temple or ancient redwood tree—lift our spirits.

The sabal palm teaches another parable. By playing the biblical fool—building its whole life on sand—it persists through impermanence. Over the life span of a sabal palm tree, often longer than a century, the landforms into which the tree germinates are transformed by the time the tree dies. This is no tragedy but the way of sandy coasts. I did not realize it at first, but the forces of the waves and moving sand have molded every part of the sabal palm's existence, from its body to its fruits, its early growth, and the chemistry of the cells in its leaves. Perhaps even the tree's name bears a sandy mark. French botanist Michel Adanson left no written record of why he coined "sabal" in 1763, but he likely derived the botanical neologism from *sable* or *sab*, sand in French and Creole.

The sabal palms on what is now the Georgia coast have had a lively evolutionary schooling in the ways of sand. For the last million years the height of the sea has surged up then down many times, driven by the alternation of ice ages and warmer interglacial periods. Ice age cycles are not so regular as the tides, but like the Moon's effects on Earth, the cooling and warming of ice ages seems to be driven by regular changes in celestial orbits. These cosmic forces are overlaid on climate variations caused by gases in the earthly atmosphere.

At their peak ice ages lock water on land, trapping it in ice sheets and glaciers. These stores are so large that the sea partly empties. When the planet warms, most or all of the ice melts back into the ocean. The meltwater, combined with the expansion of water as it warms, fills ocean basins and raises the sea. About a dozen such ice ages have come and gone. The latest ice age reached its peak twenty thousand years ago. Sea levels were 120 meters lower than they are today. In the shallow Georgia Bight, this low sea level moved the coastline east by one hundred kilometers. Any land animal with the inclination could wander dry footed as far as what is now the oceanic continental shelf.

Since the end of the last ice age, the sea has risen and the shoreline raced west, moving fast at first, then slowing. Sand spits, shoals, and islands moved with the sea's edge, continually pushed west by beetle feet and waves. When erosion was particularly vigorous, whole beaches or sandbars were lifted and dumped behind islands, creating a rolling motion of sand that tumbled barrier islands away from the leading edge of the rising ocean. The sands in which the sabal palm grows are the remains of Guale Island, which sat just to the northeast, an island that was obliterated within the last five thousand years. The island's sands piled against St. Catherines, creating dunes where previously marshland had stood. Now, as the beach continues to erode, ancient marsh mud emerges from under the disappearing sand. One such patch was exposed on the beach directly in front of the fallen palm. These dark, sticky exposures resist the waves and jut above the disappearing beach. Eventually the mud becomes seafloor as the water continues its advance.

Between the ice ages, sea levels were higher than they are today, at least six meters and perhaps as much as thirteen meters above today's levels. During interglacials, temperatures ranged from half a degree warmer than current conditions over most of the globe to five degrees warmer at the poles. This unsettled wandering of temperature and shorelines continues an old pattern. A graph of sea levels over the past five million years looks like a cross section through choppy surf. In yet deeper time, more than seventy million years ago, the height of this surf was magnificent. All of Florida and half of Georgia were shallow seas dotted with islands. The sabal palm's ancestors likely grew on the sand of these beaches and islands, with dinosaurs nibbling on their fruits.

On the scale of thousands of years, sand behaves like water. A dune is a ripple; an island is a cresting wave. The sand water rolls, churns, and streams under the power of ocean and wind. Sabal palm is a surfer on these waves, riding each one until the breaker overtops and collapses. The palm then paddles to the next rising swell, stands up, and rides the wave's face. Unlike human surfers, this rider also creates its own waves. Dunes are the result of the relationships among the biological actions of dozens of plant species with the physical forces of water and air. On a smooth beach, washed-up fronds, roots, and stems of plants cause windblown sand to break its flight and fall. These sandy accumulations further disrupt the wind, dropping yet more sand into the nascent dune. If grasses colonize the clump of sand and wrack, their roots stabilize the aggregation and may create a dune that will last for decades or, sometimes, centuries.

Beach debris from dead grasses and palms are the kernels of dunes. Living palms arrive when a seed washes ashore or is dropped by a bird. Because new sabal palm habitat is patchily distributed and may be some distance from the parent, palm trees produce thousands of fruits. Each fruit has little chance of surviving, but their prodigious numbers let the palm beat these slim odds. The fruits are about the size and color of blueberries. They bob in ocean water for months, then wash ashore, where the seeds germinate, unharmed by their soaking in salt. In the Carolinas, home to the northernmost populations of sabal palm, the seeds are particularly salt tolerant, suggesting that these palms are the descendants of ocean-borne colonists. Birds and mammals do more of the work of dispersal farther south on the U.S. coast and through the Caribbean. As robins stream along the coast in their semiannual passages up and down North America, their beaks and guts are winged freight liners for palm fruits. Year-round avian residents of the island also make regular visits to palm groves, carrying seeds to potential local nurseries. The patter of falling fruit as titmice and woodpeckers riffle through fruiting stalks is a regular accompaniment to my visits.

Once germinated, the palm must survive hardships that would defeat almost any other species of plant. These botanical challenges clash with the palms that live in our imaginations. A deck chair under a shady palm, a holiday time where cares slip away? Not so for sabal. Beaches are places of physiological affliction. Salt draws water out of roots and leaves. Heat and drought alternate with inundation by tropical storms and high tides. Blown and washed sand can bury a decades-old sapling in minutes. Lightning is common so fire often scorches vegetation. Yet the sand surfer, the sabal palm, stays on the wave.

The source of the palm's tenacity is partly revealed through the din of its leaves. When walking on fallen fronds, we hear volleys of cellulosic guns, the crackling breakage of hundreds of stiff bonds. My students pick up fronds and shake desiccated rattles to the air. Raindrops in the palm's crown are like pebbles striking a metal roof. All this sound comes from stiff silica struts within the fronds. Silica-secreting cells run with the fibers of the leaf, adding microscopic splints to the weave of plant tissues. Silica is sand. A palm frond is therefore part stone. The frond's tissues are further strengthened by a bulky surface layer of cells and by lignin, the botanical reinforcing molecule, generously woven throughout. Botanists despair when they try to cut thin slices of palm frond to examine under the microscope. The fronds' abrasive tissues ruin expensive knives and microtomes.

Each frond is supported by a lightweight, meter-long stalk, as rigid as a much heavier strut of lumber. Two straps anchor the stalk around the palm trunk. Arching from the stalk's tapered end grows a hundred-fingered hand, almost as long and wide as a person is tall, each finger stretching from a tidy pleat in the center of the leaf. From a distance, these palm leaves look like a disordered puff at the tip of the trunk, but their stalks emerge from the crown in a rosette, each stalk base precisely offset from its neighbors, like the curving pattern in a sunflower head.

Sabal palm fronds are as water thrifty as desert plants. A double layer of wax keeps salt out. Every breathing pore is sunk in a wax-capped well on the lower leaf surface, and the base of the stalk constricts water vessels so that flow is pinched. If, despite these protections, some seawater penetrates the interior, palm cells sequester the salt. They pump it into compartments within the cell, then soak the outside of these membranous chambers with chemicals designed to counteract the water-sucking draw of salt. The crack and clatter of sabal palm fronds is the sound of botanical mastery, a leaf simultaneously adapted to assault by salt and by drought.

The dunes and forests that grow alongside beaches are not always salty deserts. When rains arrive, salt washes from leaves and flushes from soil. Sand can't hold water for long, though, so palms must seize the freshwater. Thousands of worm-thick roots crawl in every direction from the swollen butts of sabal palm trunks. Like the fronds, these roots are strong, the result of many lignified fibers and sheaths. Despite their slender diameter, I cannot snap them no matter how hard I pull on roots exposed by tree falls on the beach. The roots' abundance—like a swarm of tunneling snakes—both anchors the tree and serves as water catcher. Freshwater flows up roots and into the stem. Unlike other trees, whose trunks are columns of dead tissue wrapped by skins of life, palm trunks are full of living cells. When it rains, these cells bloat themselves with water, turning the sabal palm trunk into a columnar cistern. The trunk is about half a meter wide, and every meter of height along the trunk can hold twenty-five liters of water. In dry times this store is trickled through the frond stalks' constricted bases, keeping leaves just moist enough to function. In a large palm, trunk water can keep the tree alive for months, even if the whole plant is uprooted. In a forest fire the palm's crown explodes and burns away, but water keeps the trunk alive. People who have seen these fires tell me that a burning palm grove is a polyrhythm of detonations. But within days, even if all other tree species are dead, the blackened palm crowns grow new fronds, resurrected by living tissues buried in the trunks. Old palms will persist for decades in overwashed, salty areas, riding out the waves until the very end. They continue to flower and fruit, feeding animals and launching seeds toward the sandy leavings of the ocean.

When the seed first germinates, instead of growing upward, it carves down into the sand, pushing the growing tip a meter belowground. After the initial descent, the tip swerves up. Fronds grow upward from this buried hook, poking above the surface from the burrowing trunk. This saxophone-shaped early growth lasts, on average, for sixty years. During this time the palm builds energy reserves, keeps safe from fire and sandy overwash, and expands the size of its leaf crown. Such patient expansion is a necessary preparation. Once the trunk emerges aboveground, it will not increase in girth. Unlike other trees, the living tissues in a palm trunk grow up, but not out. This different approach to growth allows palms to grow where other trees cannot, but it also forces them to invest in a prolonged juvenile period where the trunk gradually takes on its adult girth before it lengthens. For sabal palm, even this constraint can be turned to advantage. The palm can wait decades in the understory of oaks, myrtles, and other palms. When a fire or windstorm sweeps aside the shading trees, the palm launches from its well-provisioned base camp.

Changing sea levels have refined the palm's understanding of the coast, an understanding coded in its genes and its relationships with both its physical and biological companions. For the few seeds that successfully germinate and become trunked palm trees, the plant's life span often stretches beyond the century mark. Exactly how long sabal palms can live is unknown, though. Their trunks leave no "tree rings" of accumulating dead tissue. Our best estimates, though, suggest that about one hundred generations separate the sabal palm on the St. Catherines dune from the palms that grew at the end of the Ice Age, along shorelines one hundred kilometers east of the modern coast.

In the summer that followed the flattening of the dune, when the palm stood just beyond the tide's reach at the top of the beach, a loggerhead turtle visited the palm and dug a nest in its shade. She came at night and smoothed a path through the sand with the bottom of her carapace, a path marked on either side by the alternating oar marks of her flippers. The track came directly up the beach, turned under the palm, then meandered seaward. When my students and I arrived, turtle conservationists were already at work, combing through the surface of the sand to find the buried passageway to her eggs. No one saw her, but flipper marks pointed to the general vicinity of the nest. After the turtle finished digging, she shoveled sand until the hole was filled, then hid her work by turning in place and fanning the ground to blur any marks. Only by carefully slicing down through layers of beach could we find the circle of churned sand that plugged the neck of the tunnel leading to her buried nest chamber. Predatory pigs and raccoons use their noses, but humans have to puzzle over clues left in the varied texture of sands.

Once it was found, human diggers exhumed the nest, accompanied by the moan of offshore shrimping trawlers. Metal shovels peeled away sheets of wet sand until the white glow of the first egg became visible. Fingers then went to work, with arms plunged as deep as they could reach into the beach, teasing the delicate eggs from their encasement of sand. Half an hour later, 120 pearly globes, each about the size of a bantam chicken egg, nestled in the bottom of a plastic bucket. Within the hour they'd be reburied on another beach. Human labor gives these eggs a better chance. The turtle had dug her nest on a beach well patrolled by feral pigs. I've seen dozens of hogs here, digging through sand and marsh mud. A trove of 100 turtle eggs is likely to be ransacked. The rapid erosion of the beach is another danger. In the two months that it takes for the eggs to mature and hatch, the beach will move inland by a few centimeters, perhaps more. Even without such shoreline movement, high tides on an erosion-flattened beach can drown a turtle nest. The beach to which the conservationists moved the eggs is kept clear of pigs and is one of the only spots on the island that accretes sand. This egg nursery buys time for the loggerheads that lay on St. Catherines. Two decades ago one quarter of the shoreline was suitable for turtle nesting. Now erosion has more than halved that proportion.

Sea turtles have an even longer genetic memory of the coast than palms. They have been crawling onto beaches and digging nesting holes for over 100 million years. Yet all seven modern species of sea turtle are now imperiled. Adult turtles are killed by boats and nets. Erosion and development have squeezed their breeding grounds. Predators, including humans seeking the reputed aphrodisiac properties of raw turtle egg, are abundant on the remaining beaches. Turtle conservation programs, like the one on St. Catherines Island, therefore shield the brief terrestrial life stage of turtles from a shore that is arguably more unwelcoming than at any other time in their history.

For humans who labor on behalf of turtles, the scrape of tiny flippers on a beach as hatchlings rush from their nest to the sea is perhaps the sweetest sound that can pass under a sabal palm's fronds. More bittersweet is the splash and gurgle of the swimmers as they scull straight into the Atlantic. Gulls loiter offshore, waiting to pluck morsels from the swell. Beyond the gauntlet of predatory seabirds, thirty years of life in the ocean separate the hatchlings from their first return to land as breeders. Many of these young turtles will follow the Atlantic gyre, passing Iceland, northern Europe, and the Azores and finally arriving in the Sargasso Sea, just beyond the Georgia Bight, where they will live until they reach sexual maturity. Some will avoid the swirl of the Atlantic and swim directly to the Sargasso Sea. One in a thousand will survive to breed. When mature, the females come back to shore to lay their eggs; the males never again set foot on land. Like the sabal palm, theirs are lives that draw our imaginations into the future of the sea and its coasts.

As tides recede from the roots of the sabal palm, they leave rafts of sea-foam. These clouds seldom reach taller than knee height, but they can be as long as a small boat. The bubble rafts are surprisingly firm and long-lived. The wind lifts them from the surface of the beach, then drops them meters away, unharmed. When a skin of water covers the beach, the rafts creep like snails, moving over the slick surface under the power of the breeze. I scoop some foam into my hands and thousands of bubble surfaces pop as I lift them, a sizzle like fish frying in a pan. The smell of foam is a distillation of ocean, like an inhalation after diving through a collapsing breaker, head wet with spray.

The foam is made from the pulverized remains of algae and other microscopic life. When these cells break apart in the tumult of the ocean, they release proteins and fats into the water. These chemicals act like soap in a bath, changing the surface tension of water. When the wind agitates the water surface, like a hand whipping a bubble bath, the result is a froth. Sea-foam is a memory of the biology of the ocean, blown to land. Seawater isn't only water but a living community. Turtles are one of the more obvious and charismatic members of this assemblage, but they do not represent the majority of the ocean's life. A single drop of seawater contains from hundreds of thousands to tens of millions of microbial cells.

The life of this community is like the network in a ceibo crown or a balsam fir root, but liberated from the constraints of stationary terrestrial life. Ocean microbes mix freely and their watery surroundings allow cells to exchange chemicals without complicated bonding and attachments. Water has further dissolved atomism, reaching all the way into the DNA of ocean microbes. A balsam fir root communicates with the DNA of the other species around it, but the root retains all its own genes, as do the other partners in the conversation. In the ocean, though, microbes take interdependence a step further.

Each microbial species in the ocean specializes in a particular task—gathering sunlight or reworking organic molecules—and abandons most other tasks to the community. Evolution has weeded the DNA of these species, leaving each one only the genes needed for its specialized task. Other tasks, even though they are central to the cells' lives, are performed by other microbes. This "streamlining," whereby individual species lose genes for essential tasks and come to depend on the community, is possible only because the microbes float in proximity to one another, letting chemicals move from one cell to another with ease. Some cells exchange not just foodstuffs but information. Molecules signal need and identity, allowing specificity of exchange among cells, even in the ocean's turbulence. If separated from their community, many cells die. Their DNA cannot meet basic needs.

The smallest viable genetic unit of microbial life in the ocean is therefore the networked community. This arrangement is efficient, allowing each part of the network to focus on what it does best, but it is vulnerable to disruptions in communication. If relationships among cells are broken by oil spills, synthetic chemicals, or changed ocean acidity, the microbial community transforms, with consequences that reach beyond microbes. The atmosphere and ocean's chemistry depend on these networks: Half of the world's photosynthesis happens among ocean microbes and plankton. Billions of whispers among ocean creatures therefore determine the chemical state of the world's air and water.

We do not know how changes to the ocean are disrupting the exchange of information among cells. The networked genetic streamlining of the ocean has been discovered only in the last decade. But long-term surveys of the ocean suggest that planktonic life has declined at an average rate of 1 percent per year for the last century. Fish populations are in many places in precipitous decline. The chemical nature of the ocean is also in flux. Acidity is increasing as carbon dioxide dissolves into water, and novel chemical products from humans wash downstream and float in every drop of ocean. Some of these chemicals disturb or break communication among cells in human bodies. They may do the same within cellular networks in the ocean.

Within hours of the sabal palm falling, the blowing spume brought another novelty, a shard of white plastic. The plastic embedded itself in the overlapping frond bases that, until the tree fell, had been the homes of lizards, frogs, and ants. The plastic is one of tens of thousands of fragments that lie around the tree. The skittering of plastic bottles across the beach and the thrash of plastic sheets caught in tree branches are a regular part of the sabal palm's sonic environment. With my students I have surveyed this washed-up refuse around the tree. We used linear surveys of standardized length, measuring every visible fragment in the wrack line. If our samples are representative of the island, a ten-kilometer stretch of beach holds just shy of half a million pieces of visible plastic. We did not dig below the beach surface, so more plastic is likely present. Small pieces far outnumber larger fragments; half of the pieces that we gathered measured less than two centimeters across. Studies of other beaches show that this trend continues under the microscope: the smaller the fragment size, the more pieces are found. As living plankton dwindle, planktonic plastic takes its place.

Thoreau also left a record of his beachcombing for the "waste and wrecks of human art." His gleanings and those of my students are protoarchaeology, glances at cultural artifacts from two times.

Unmeasured saunters along Cape Cod shore, 1849, 1850, 1855

> Logs washed from the land (many)
> Wrecked boat lumber and spars (abundant)
> Pebbles of brick (a few)
> Castile soap bars (not counted)
> Sand-filled gloves (one pair)

Rags and tow-cloth pieces (not counted)

Arrow head (one)

Water-soaked nutmegs (boatload)

Items in fish stomachs: snuff boxes, knives, church
    membership cards, "jugs, and jewels, and Jonah"

Box or barrel (one)

Cord, buoy, piece of a seine (one)

Bottle, half full of ale that "still smacked of
    juniper" (one)

Barrels of apples (twenty, second-hand report)

Human cadavers (at least twenty-nine)

*St. Catherines Island wrack line, 2013–14, survey lines
covering 160 square meters*

Blocks of buoyant plastic foam (one hundred and
    sixty-three)

Plastic drink bottles (twelve)

Plastic pill bottle (one)

Balloons, crinkly plastic, deflated, Happy Birth-
    day (two)

Balloons, rubbery plastic, air-filled, Just Married
    (one)

Air-filled latex glove, trapped under palm log (one)

Plastic two-gallon juice jug with seventy-five
    barnacles attached (one)

Blue plastic bucket with label stating, *Verwijderd
houden, Gas niet inademen*, Readers of Dutch,
Keep away, Do not breathe gas (one)

Black plastic bucket with "T1 Heavy Duty Engine
    Oil" on label (one)

Bottle caps, plastic (two)

Woven plastic ribbon, purple (one)

Laundry tub, white plastic (one)

Flip-flop shoes, plastic, unmatched (two)

Jar of mayonnaise, plastic, half full, "still smacked of" emulsified vegetable oil (one)

Plastic bottle of tobacco chewer's spit, smack untested (one)

Fishing buoy, plastic, with rope (one)

Shotgun shell, red plastic (one)

Casing from rifle bullet, brass (one)

Fragments of hard plastic of varied colors and shapes (forty-two)

Rubbery inner globe of bald tennis ball (one)

Items in stomachs of necropsied turtles, beached animals killed by propeller strikes and starvation: metal fishhooks, jellyfish-size transparent plastic bags (two)

Lengths of plastic rope (three)

Boards of pressure-treated wood (five)

Glass bottles (two)

Rusted ship's ladder, serviceable (one)

Metal spray can of Essence of Man perfume (one)

Archaeologists use their findings to infer the customs, productions, and religious beliefs of the cultures that they unearth. At the sabal palm's roots, we have evidence of a civilization that first left wood and glass, then underwent a plastic revolution, all in a few decades. The manufacture and movement of plastic is the oceanic detrital signature of our age. A creative archaeologist might deduce religious significance from the flotsam: some artifacts are food and tobacco offerings, and there is unambiguous textual evidence of plastic at the center of rituals that mark conjugal and aging ceremonies.

Floating plastics change the ocean's living networks. The intestinal tracts of animals from turtles to birds to sea worms are choked, lacerated, or slowed by the plasti-plankton. Less obvious, but of greater importance to the ocean's cycles of energy, life, and matter, are the effects of billions of microscopic fragments of plastic. The tempo and mode of microbial life in the ocean emerges from the exchange of chemicals among free-floating cells. A haze of plastic particles is a novelty that reworks these microbial relationships. The specks of plastic are hard surfaces on which cells aggregate in communities seldom before encountered in the ocean. Some microbial species grow only on solid surfaces. Formerly these were rarities in the open ocean. Now they are common. Plastic shards serve as islands on which rare species that seldom bumped into one another are held in proximity.

A few of the colonizing microbes bore into the plastic surface with digestive chemicals. As these pits grow, the plastic breaks apart. Breakage, and the weight of the life-forms that smother each fragment, sink the plastic. Although we understand the process poorly, microbes seem to remove plastic from the surface of the ocean. Some fragments sink; some return to their original chemical constituents. In an imperfect way, modern microbes may be finishing the job of their ancestors. Oil comes into being because microbes do not fully digest the dead remains of algae and plants. Instead, geology takes over and turns dead greenery into liquid fossil. Humans now turn this fossil to plastic that, after a single use as a bottle or bucket, goes to a landfill or the ocean. Microbes may close the loop, although their work is not fleet enough to spare turtles and worms.

As plastic reworks the ocean's life, the shoreline's movement continues. In the nineteenth century, sea levels rose across the globe by an average of just over a millimeter each year. In the last two decades, this yearly rise has accelerated to three millimeters. Ninety percent of the extra heat that we've added to the world in the last decades has been absorbed by the oceans and swept to its depths. Like liquid in a thermometer, this heated water takes more space. All our projections of the next decades predict that more heat will be pumped to the ocean. Melting ice sheets and glaciers also add to the ocean's volume and these ice losses are accelerating. We don't know exactly how much thermal expansion and meltwater to expect, but some robust, well-reasoned studies suggest that by 2100 seas may rise by somewhere between one half meter and two meters. Other more conjectural studies suggest an even greater rise.

From the perspective of a sabal palm or a sea turtle, these are modest changes, well within the range of experience of their recent ancestors. But the days of surfing sand waves are temporarily gone. Dunes cannot flow across marshland; islands don't flip and drift inland, forming new sandy waves. Instead, these geologic processes meet walls. We have replaced maritime forests that held nurseries of young trees with roads and towns. High tides wash against piles of riprap stone imported from inland and lap at stone that fronts buildings and asphalt. Sand that formerly flowed into the ocean from rivers is held upstream, trapped by dams. The longshore currents that replenished beaches are now sand starved, so erosion continues, but accretion is halted. All exhale, no inhale: the beach withers. Eventually the ocean will bury all this hardscape and obliterate humanity's attempts to impose changelessness. In the meantime the plants and animals of the coast are colliding with shoals for which their previous lives had not prepared them.

If projections of sea-level rise are correct—and so far the sea has outstripped past projections of climate models—then human calamity is likely. Ocean water may claim the land on which are built the homes of over 2 percent of the world's population. Storm surges and crumbling shores will affect the 600 million others who live within

ten vertical meters of sea level. Within the next two human generations, the sea is likely to turn many people into evacuees.

Beached cadavers are the part of Thoreau's seaside wandering
that seems most discordant with modern experience on the U.S.
coast. On October 7, 1849, two days before Thoreau arrived on Cape
Cod, a brig from Galway, Ireland, slipped its anchor in a storm and
wrecked, drowning many migrants. Thoreau was sanguine, removed
from the funereal scene. He "sympathized rather with the winds
and waves," believing that the Irish had "emigrated to a newer world"
where they kissed the shore in rapture, leaving their bodies to toss in
the surf. Thoreau sounds callous to modern ears. His ambivalent
opinions about the worth of the Irish perhaps distanced him
from their fates. He was also inured by the scale of immigration in
his time and the frequency of wrecks. In the Great Irish Famine
more than a million men and women fled to the United States, a
country that had just over twenty million censused inhabitants in
1850. In Thoreau's time a ship wrecked on Cape Cod every fortnight
through the stormy winter. "Why," Thoreau asked, "waste any time
in awe or pity?"

We found no human bodies under the sabal palm, although
plastic-killed turtles or birds sometimes wash ashore. Our coasts
seem far removed from Thoreau's. But they are not. It is no longer the
meanness of potato disease and nineteenth-century English politicians that drive the upsurge in refugees but new dislocations, the
rising sea included. The numbers are controversial; quantifying migration in response to environmental change is an inexact study at
best. But most estimates claim tens of millions of people have so far
been displaced by the effects of moving coastlines, degrading soils,
and dwindling freshwater, with perhaps hundreds of millions more
to come. The Georgia Bight sees little of this movement. But on the
shores of the Mediterranean, the Gulf of Aden, the Andaman Sea,

and the Canary Islands, tourists once again wander past the bodies of migrants and wreck survivors crawl past lounge chairs on the beach. Twenty-first-century English politicians echo their ancestors: "We do not support planned search and rescue operations in the Mediterranean. We believe that they create an unintended 'pull factor,' encouraging more migrants." Thoreau's world of mass migrations and shipwrecks is back.

Like the gravitational forces that link the orbits of Sun, Earth, and Moon, the physical laws that connect temperature and the properties of water are not negotiable. For every one hundred liters of melting Antarctic ice, ninety-one liters of water flow to the ocean. One degree of warming increases the volume of tropical ocean water by three one hundredths of a percent. Sabal palm has not fought the rules. Rather, evolution has found creative ways for the sabal palm species to live in the interstices among the Newtonian forces that act on sand, salt, and tide.

We're not palms, able to jump among beaches through wave-bobbing and bird-winged seeds. But as the ocean unsettles the old human order, we might attend to the trees, not to mimic them but to better understand life on the shore. Sabal palm has learned to thrive amid changeability. It outcompetes tree species whose ways of being are better suited to inland mountains and plains. For most of its life, sabal palm grips the sand, resisting the action of waves, holding dunes in place. The tree desalinates its cells and contrives to store as much freshwater as it can. When storms and fires arrive, the palm accommodates, then rebuilds. Eventually, though, fronds and roots, once so crisp, are entrained and drowned in the push and suck of surf. The sabal palm moves on, leaving a boneyard of timber on the beach.

The biblical parable can be reworked. The fool is not the person who builds on sand. The error is to believe that sand can be rock. No matter how much concrete we pour, we can never turn the coast to stone. Instead, the wise build on sand knowing that its nature requires both creative resistance and the ability to walk away. Human society has so far emphasized resistance but given little help to those who by choice or misfortune must take the second path. "Why waste any time in awe or pity?" Perhaps because our answer to the sea lies in what the palms lack: networks of mutual aid.

Green Ash

Shakerag Hollow, Cumberland Plateau, Tennessee
35°12'52.1" N, 85°54'29.3" W

There is life after death, but it is not eternal. Death does not end the networked nature of trees. As they rot away, dead logs, branches, and roots become focal points for thousands of relationships. At least half of the other species in the forest find food or home in or on the recumbent bodies of fallen trees.

In the tropics, soft-wooded trees pyre their bodies in rapid, smokeless blazes of bacteria, fungi, and insects. Their fallen logs seldom last longer than a decade. Tropical trees with denser, heavier wood linger for a half century at most. The process of decay takes much longer in the acid cold of a near-Arctic bog. There a tree measures the river of its afterlife in spoonfuls fed to patient microbes over millennia. Between the extremes of tropics and poles, in the midlatitudes, a downed tree in a temperate forest might live in death as long as it stood in life.

Before its fall, a tree is a being that catalyzes and regulates

conversations in and around its body. Death ends the active management of these connections. Root cells no longer send signals to the
DNA of bacteria, leaves end their chemical chatter with insects,
and fungi receive no more messages from their host. But a tree
never fully controlled these connections; in life the tree was only
one part of its network. Death decenters the tree's life but does not
end it.

In Tennessee springtime brings collisions between walls of Arctic air
and pushy bubbles of warm moisture from the Gulf of Mexico.
Windstorms ensue. Gusts and bursts from the sky unmask any
weakness in trunks or roots. On one such wind-bruised day I was
wandering across a wooded, rocky mountainside and came on a giant green ash tree immediately after it fell.

March

Their every footfall is a dry splinter of sound. Six thousand chitinous
feet raise a shudder of air, a scrabble of bark. The insects wrestle, they
mate. Slaps punctuate their writhing as clumps drop to the leaf litter.
Grounded, on they fight, then break apart and fly arcs to the tree,
wings *zizz*ing. Wasped in black and yellow, with tendrilous antennae, they are fearless as I approach. Their dissembling countenance
protects them: even though the insects are beetles, the color of their
bodies, their confident behavior, and the sound of their wings are
hornetlike.

   These insects are banded ash borer beetles, here only for a day.
They mate, then till eggs into the ash's bark. This morning even
this dull-snouted human could find the wind-felled tree. It loosed a
scent of tannic acids, sour and opaque as oak, with a whiff of brown

sugar. Now, hours after the fall, only the tang of bruised bark remains. The beetles live by this reek of rent wood. Freshly downed ash trees are their nursery. Tucked under bark, larvae auger all spring and summer with jut-bladed mouths. The larvae swallow the fine sawdust, passing it to guts populated by symbiotic microbes. Without these wood-digesting companions, the beetles could not feed.

I bend my ear to the log and hear clicks, rasps under the spongy bark.

April

A buckeye sapling grew at the green ash tree's foot. When wind felled the ash, the youngster took a ride, swinging up by a meter and swiveling ninety degrees. Now buckeye buds, shuttered since late summer, drop their scales and wake to a perpendicular world. Gravity comes to them sidelong.

The primordia of the year's leaves swell in the bud, cellular knots fanning to reveal a leaf in miniature. Inside the cells of the opening buds, descendants of ancient bacteria sense gravity's changed direction. These bacteria, now transformed into saclike amyloplasts, have traveled inside plant cells for 1.5 billion years. Amyloplasts now serve as the pantries of plant cells, storing starch. When gravity shifts, these starch dumplings roll and sag, pulling on amyloplast membranes and sending a signal to the rest of the leaf. *Cells on the underside of the stem: elongate. Cells on top: hold steady.* The stem rights itself, a curve that turns the expanding fingered leaves Sun-ward.

The buckeye's growing tip now points directly up. Such perfection of sensation and response depends on signals among the many creatures that compose a plant cell. If the amyloplasts are faulty or if the rest of the cell is deaf to their signals, the stem can feel no gravity.

May

Forty-one meters from the torn roots, the ash's crown is now a mess of shattered limbs. The poke and tangle of twigs reaches to my eyes, unwalkable. But crevice and ravel of wood beckon to Carolina wrens. A pair plunged into the thicket one day after the fall, looping their song through the wooden snarl. Now their territory is centered here; they chirrup and rasp in call-and-response on my every visit. The birds wing and dive under branches, delivering beak-snared gnats to their nestlings. Throughout the forest, in every downed treetop, russet feathers of wren pairs slide through mazed branches. They are bird mammals, hole seekers, wreck chasers, their lives made possible by the protective tangles of smashed tree limbs.

June

The tower of sunshine admitted by the downed tree brings heat to the subcanopy and leaf litter. Forest animals know where these basking spots lie. All around the litter is dusk; here the light is bright and hot. On a warbler-sung morning I sit in the sun patch, watching. After an hour or more, a thumb-size tawny curve in the litter rises into my consciousness. My eyes suddenly open wide. I see segments, armor plates, then my breath snags: rattlesnake! Two boot lengths away from the unseeing fool enthroned on his log, the snake sleeps. It gives no cicada-in-dry-leaves warning. An arm-thick body curls on itself; head and rattle kiss. Looking like sun-bleached dead maple leaves mixed with soil-dark oak, the snake has perfect lying skin.

I peer closer and see that the rattlesnake's eyes are clouded. Perhaps the snake is preparing for a skin molt; hazing of the eyes is normal before the animal sloughs its skin. But fungal infection is another possibility. All animals have fungi on our skins, mostly harmless commensals. But in the last five years, fungal communities on the skins of rattlesnakes across the eastern and midwestern United States have changed. One fungus species has overwhelmed the others, sickening or killing the snakes. The cause of this change is, for now, obscure. Warmer winters may make one fungus dominate the others, or trade in foreign pet snakes may have imported a new, more aggressive fungal strain.

Whatever the cause, rattlesnake skin disease is spreading, with unknown consequences. Rattlesnakes are directly and indirectly connected to many other species in the forest. Their prey, rodents, are predators and dispersers of seeds and nuts. Depending on how rodent populations respond to the loss of rattlesnakes, the fate of tree seeds may change, perhaps suffering heavier loss to rodent appetites. Rodents are also the main vectors of tick-borne diseases, so fewer snakes may result in more parasites inside the blood of birds and mammals, humans included. Owls and hawks feed on some of the same food as rattlesnakes, as do foxes and coyotes. Our understanding of these linkages in the forest network is too imprecise to predict how the numbers and behaviors of these other predators will change as disease spreads through the snake population.

I return the next day and the snake continues its sit, coils shifted slightly, eyes still clouded. Two days later the rattlesnake is gone, leaving a hand-size depression in the leaf litter.

August

A fly perches on a sassafras leaf, washing its pollen-fuzzed forelegs. Seeing me, the gold-banded insect flicks into the air. The leaf bobs as the legs push away. The flight is too quick to follow, but the wing thrum is as loud as a hive. Like the ash borer beetles, the fly protects itself by looking and sounding like a stinging insect.

"News bee," we call it. It finds my face, lines up with my nose, and delivers its story eye to eye. The fly wavers, weaves a little, causing the beelike whine to quaver. Then it zips away, covering several meters in a second, and tells the same story to a stain on a maple trunk before flinging itself away across the mountainside. The bee is an active hunter with large eyes that search out the forest's late-summer flowers. My face and the maple trunk are interesting enough for a seconds-long look but, lacking nectar, we do not hold the bee's attention for long.

I'd met this species before, in the ash's split trunk. Water collected in the rotten interior hollow, like a swamped wooden canoe. At first the water was the color of honey. Days later it was well-steeped tea, then, after weeks, clouded broth. On this cloud the insect larvae arrived, swimming in their soupy food: wormy gnats, twitching mosquitoes, and crawling grubs. As the larvae grew, some of them dimpled the water surface with the tips of their breathing tubes. Each tube was as wide as a hair, connecting the crawling aquatic grub to the air. These divers and their air tubes were rat-tailed maggots, the young of the gold-banded news teller, each maggot feeding in the trunk puddle's detritus.

There are no ponds or lakes in this forest, but tree holes and split trunks are like bromeliads on a tropical ceibo, holding water in which hundreds of species thrive. Deadwood becomes the forest's aquatic habitat. The "news" is: every knothole, cracked log, or wad of old leaf is a pool or bog. Gnat-hungry nestlings, pollinator-yearning flowers, and every creature bitten by a mosquito is connected to the dead tree's swamp.

October

Something—perhaps the promise of a view from a meter-thick log or the sturdy foot-gripping bark—attracts mammals to the walkway.

By day the log is seldom lacking a chipmunk or squirrel. At night coyotes, squirrels, opossums, woodchucks, even a well-muscled bobcat use the highline. Predators sit and look out over the slope. All others amble, free of tick-brushing vegetation. Nightly a raccoon family pads in a file of four, running the whole length of the ash trunk. In the evening they walk from crown to root-ball, headed upslope, perhaps to the town trash containers that lie half an hour's walk in that direction. Later at night they return and walk root-ball to crown, then down the forest slope, presumably to a den somewhere in the jumbled rocks below.

The walkers leave their scat, territorial markings perched atop the log. Each deposit is a record of the animal's diet, a glimpse into the food web. One such wet, thumb-size dropping is a well-packed mélange of cricket legs, wasp heads, leaf pulp, seed, bee abdomens, and a single horsehair worm as long as my hand. I probe another, a finger of fox dung, with forceps improvised from fallen maple twigs. The whole sausage is made from wild grape seeds, glued with tar. I pocket two seeds and plant them in a windowsill pot. When they germinate, the blush of white wax under their leaves identifies them as pigeon grapes. The name echoes the passenger pigeon, now extinct. When the pigeons gathered here, they carried all manner of plant seed, depositing it in carpets of guano. Like balsam fir, plants here depended on animals to carry their young. Now the courier is gone. Fox and raccoon must do the botanical work of billions of pigeons.

Between scat piles, flakes of glowing orange lie in the bark crevices. These plastic fragments are the gnawed remains of a golf ball, the plaything of a human who knocked the ball into the woods. On arrival the bright globe became the chew toy of a creature with strong teeth, perhaps a playful coyote pup or a confused squirrel.

In drawing mammals to one place in the forest, the log draws the forest's future around itself. Seeds accumulate in piles of fertilizer.

The log's decay will drip more nutrients to ground, mulching and feeding germinated plants. As in the seawater around the sabal palm's roots, shards of plastic garland the log, industrial forest flotsam.

November

I withdraw a long-stemmed pipette from the murky water pooled inside the fractured log. Squeezing the bulb, I then ease a single bead of log liquor onto a microscope slide and flatten the water drop with a slip of fine glass.

At 40x magnification, I see midge legs and shredded wood. I grasp the microscope's knurled ring and twist more powerful lenses into place. At 100x drunken needles of light move across my field of vision, wood shards caught in the swirl of water. At 400x the microscope's eyepiece is full of the jostling, shuddering motion of living cells. In one drop there are more creatures than there are trees on this mountainside. Double balls track ruled lines, swimming back and forth across the slides, nodding as they flow. Comma-shaped cells snake and cruise. Jelly dolls rotate. Slipper-shaped cells creep, water swirling in their wakes. A giant—fifty times as long as the prey in its translucent belly—hurtles across and out of sight. I crank the dials, skating the slide under the lens, trying to keep the monster in view. Water haloes its ovoid body, light bending as its rays graze the pelt of beating hair.

Three hundred years ago van Leeuwenhoek polished glass into lens. For his reports of animalcules and cockles, he was mocked as a sot by the Royal Society. Even now these microscopic creatures live in encyclopedic mute darkness; their conversations have fewer human eavesdroppers than the chatter that falls from the stars.

December

A *Narceus* millipede weaves through bark crevices, slinking its buffed leather body between long blocks of stacked cork. The millipede keeps its rounded head lowered as it moves, grazing on runnels of algae. Just as the log is habitat for the millipede, the millipede is habitat for others. Two tawny mites cling to the millipede's back, each only a tenth of the width of a single millipede segment. As the millipede chews on the log bark's rotting surface, mites sup on exudations from the millipede's exoskeleton. This is an old relationship, cemented into evolutionary genealogy. Every member of this taxonomic family of mites, the Heterozerconidae, lives exclusively on millipedes.

The life cycles of the two parties are synchronized. When millipedes retreat under logs in late summer to breed, mites dismount the host and find their own mates. In early autumn the young of both the mite and the millipede hatch from eggs kept moist under the dead tree. Where there is no deadwood to shelter nurseries, both are absent from the forest.

January

The ash's upturned root-ball stands as tall as my head. Severed roots, some as thick as human thighs, jut from clumps of clay. The winter frost has clamped windblown tree seeds onto this bare soil, holding them exposed on the surface, revealing what will germinate here in the spring.

The V wings of sugar maple helicopters are as tattered as molting vultures, torn and abraded by several months' worth of wind, rain, and now ice. Red elm's fruit is a slash of half-rotted paper, a lentil-like seed inside. The dagger of tulip tree fruits have experienced no such decay, having fallen from their tight clusters in the canopy just this week. Each fluted blade ends in an upturned nub that contains the seed. Among these native plants lies the seed of an immigrant species, a crispy plume, bulging at its center: *Ailanthus*, originally from East Asia.

When roots emerge from these seeds, *Ailanthus* will ooze into the soil a chemical that poisons all roots but its own. The seedling's roots will also converse with soil microbes in a language alien to native trees, encouraging bacteria to squeeze more nitrogen from the soil. Thus fertilized, the stem of *Ailanthus* will surge upward, shading its competition. By reworking the soil community, eliminating some connections and strengthening others, the tree has become one of the most common early colonists of gaps in the deciduous forests of the United States.

February

The bill for dinner:

Purr of sore-throated cat	Downy woodpecker
Marching-band drummer, too stoned to care	Hairy woodpecker
Tattoo in trios, quads	Red-bellied woodpecker
Wooden ruler twang, then haphazard knocks	Red-headed woodpecker
Old man in no hurry to nail loose boards	Pileated woodpecker

Racing heart fades to stammering palpitation	Yellow-bellied sapsucker
1890s telephone ringer, wooden edition	Northern flicker
Lead pellets hurled at tree	White-breasted nuthatch
Forceps jabbing leather	Carolina chickadee
Screwdriver jabbing leather	Tufted titmouse
Mechanical pencil lead thrust, twisted, snapped between book pages. Unstopping furious edit.	Brown creeper

Such are the vibratory anxieties of beetle grubs, carpenter ants, and bark snails. I lean my head to the trunk, twenty meters away from the drilling woodpecker. Inside, the tap sounds as if it were directly under my bark-pressed ear. The bird's tremors inhabit the wood, wrapping every insect in the presence of its predator.

These birds are bark-crevice experts. With flickering tongues and lancing beaks they master the tree's beetling crags. As they move over the tree, they turn their ears to what lies below, listening for any sound that betrays an insect's presence in the log.

Beetle and bird alike feed from the ash's deadwood. The sunlight stored in the ash's cellulose flows first into beetle flesh, then into the gizzards of birds. The sound of yammering impacts of beaks, driven by knot-muscled necks, is the ash's energy, part of the tree's long, final exhale into the forest.

First Birthday

The tear in the canopy caused by the tree's fall created an inverted geyser of light. After a year under the abundant flow of photons, what were small herbs have turned into bushes. Elsewhere in the forest baneberry, blue cohosh, and waterleaf brush my ankles as they sip on meager, shaded light. Here the plants have such well-provisioned roots and stems that I must lift my arms to walk through them. They are a choke of green; my every step releases the aroma of bruised leaf. I cannot see the forest floor for the profusion of plants, so my feet move cautiously, remembering the rattlesnake.

Young trees and shrubs also surge into the light. White-tailed deer have adopted as a favorite this patch around the fallen ash. They bed among the upstart spicebush, elm, *Ailanthus*, and buckeye saplings, each plant an arrogance of gorgeous, uplifted limbs. Nearly every time I visit now, I evoke a snorting alarm, chasing deer from their convivium of food and shelter.

One hundred meters away, the ash has a fallen companion, a white oak, wide as a grand door, wind-broken from its slashing fall. Its roots flung soil twenty meters away and left what looks like a bombed crater. Another huge tree, a sugar maple, snapped in the same spring windstorm. Most of its trunk was hurled downslope, leaving vertical shards attached to the untorn roots. Each pencil-thin standing splinter is taller than my reach. I grate my hands on the rent wood, plucking each piece to release a low, muffled *oong*. Like the ash, both these tall trees opened holes in the canopy. Both logs were laid open by the storm. Forest life will now move in and feast. Seen from high above, such tree falls would look like spots appearing randomly across the forest, each one a nexus for forest life. Over the years, each spot's life is gradually suffused into the surroundings.

The accountants of forests say that the world's forests contain 73 petagrams (73 thousand million metric tons) of downed trees, one fifth of the world's wood. And wood that has eased into the dead organic matter of soil, no longer recognizable as trees, outweighs the living.

Second Birthday

Hundreds of protrusions as thin and long as toothpicks bristle over the ash's surface. Each emerges from an awl hole in the bark's fissures. I touch one and it disintegrates into powder, like ash from a burned-down cigarette. I press my ear to the tree but hear nothing. I try a stethoscope, a microphone, but these fine tubes of sawdust are coming from woodwork too deep, made by animals too small mouthed for sound to penetrate to the surface of the bark.

The crumbly protrusions are the work of a tree-mining insect, an ambrosia beetle whose body is about half the size of a sesame seed. When the beetles arrived at the tree, they brought help. Inside pockets on every beetle thorax, the colonists carried wads of fungus and bacteria, companions that transform wood into ambrosial beetle food.

Unlike the bark ash borers and other bark beetles that feed on the soft layer of tissue between bark and wood, these ambrosia beetles dig deep into the log. They tunnel pinholes all the way to the core, emptying their pockets of fungi and bacteria as they go. The beetles are farmers. Tunnels are their plowed furrows; fungi and bacteria are their goats, sheep, and cows. These microscopic livestock graze on the wood, digesting it and storing its nutriment in their bodies. The beetles then return and harvest their "meat," eating much of the fungal and bacterial growth but leaving enough to continue the work of disassembling the log. This troglodytic ranch is older by sixty million years than any human farm, old enough to result in complete mutual dependence. If their bond is severed, ambrosia fungus and beetle die.

Other fungal species followed the beetle pioneers into the wood, using the tunnels to invade the log from within. This week several of these fungi shoved their heads out of the log and up through the bark. The first is a gleam, a wet brown bud with a creamy fringe, a turkey tail fungus. Another is a flat crescent of buttery yellow. Adjacent is a shelf striped in fawn, walnut, and down-feather white. These shelves and bulbs are the spore-producing outer manifestation of fungus bodies that ramify through the wood. Each spore is one thousandth the weight of a human cell, so a wind gust, a passing insect, or a tumble of twig will suffice as a mechanism of escape. Then, if a spore arrives on a hospitable wooden surface, the propagule hatches and burrows. The growing fungus worms down into this new home, using digestive enzymes to cleave wood into its sugary chemical constituents.

Within weeks the turkey tails and other fungi on the ash will themselves be galleried with crawling, feeding creatures. The young of hundreds of insect species live within fungal shelves and decaying wood: beetles, moths, and flies for which wood-rotting fungus is the only home.

When a being—a person, a tree, a chickadee—full of memory, conversation, and connection dies, the network of life loses a hub of intelligence and life. For those closely linked to the deceased, the loss is acute. An ecological analog of grief unfolds in the forest: for the other creatures that depend on living trees, death ends the relationship that gave them life. The living tree's partners and foes must all find a new live tree or they will themselves die. Much of the understanding of the forest that dwells embedded in these relationships also passes away. The trees' particular knowledge of the nature of light, water, wind, and living communities, gained through a lifetime of interaction in one location in the forest, dissolves.

Yet by catalyzing new life in and around their bodies, dead trees bring about new connections and thus new life. This creative process is not didactic or preceptive. The tree does not pass on what it knew, re-creating a new version of itself. Rather, around and inside the tree, death brings about thousands of interactions, each one exploring ecological opportunity. From this unmanaged, uncontrolled multitude, the next forest emerges, composed of new knowledge embedded in new relationships. Like a lightning rod, the dead tree draws into its body the potential that surrounds it, focusing and intensifying what was diffuse. But unlike lightning, this surge does not flow into the ground and disappear. Life instead feeds on the closeness of connections in the dead tree, increasing in vigor and diversity of expression.

Our language does a poor job of recognizing this afterlife of trees. Rot, decomposition, punk, duff, deadwood: these are slack words for so vital a process. Rot is detonation of possibility. *Decomposition* is renewed composition by living communities. Duff and punk are smelters for new life. Deadwood is effervescent creativity, regenerating as its "self" degenerates into the network.

Interlude: Mitsumata

Echizen, Japan
35°54'24.5" N, 136°15'12.0" E

L inguistic troubles delayed my pilgrimage. A map and well-rehearsed Japanese phrases did me little good at the train station's taxi stand. The shrine was not distant, but I stirred confusion when I spoke the word, 神—*kami*, the deity. Only when I stepped back, clapped twice, and bowed as if at a temple did the driver's brow unfurrow and a smile emerge. We sped through rice paddies, aiming at the hills. He dropped me on the threshold of the Paper Goddess's home, at the toe of the slope. The price of this last leg of the journey was three sheets of pulped mitsumata bark and abaca leaf fiber. On the paper's surface: a bacteriologist's portrait, Mount Fuji, and cherry blossoms. Unlike spongy Andrew Jackson greenbacks, tough rags of cotton and linen, these were bright notes, holding their snap even though they'd been played many times.

I passed through the *torii* gate onto gold-leafed flagstones, spent ginkgo softening footfall to the kami's shrines. Before I could reach the paper deity, though, I stopped at mountain water, at a pool of cool ablution. The craft of papermaking takes the same route: we cannot make paper without first dipping into cold water. Every shrine has water for purification, but here, at the home of Kawakami Gozen, cold water from the hills is also the source and continuance of the paper *kami*'s presence. When Kawakami Gozen was asked where she came from, she answered only, "From above the river." The calligraphy on the *torii* gate etches water and mountains in her shrines' names: 大瀧, Otaki, the great (大) waterfall (瀧); 岡本, Okamoto, the hill (岡) book (本). Both shrines predate her arrival, so Kawakami Gozen emerged in a watery confluence of history. Her knowledge too is an old story, coming from China via Korea, carried to Japan with Buddhism in the seventh century.

In a bath of mashed inner bark of kozo or mitsumata trees, water floats the plant cells' dissociated strands, buoying them as they drift. Each molecule of cellulose is a strand of sugars, up to fifteen thousand links in the chain. Suspended in a haze of water and hibiscus mucilage, they crosshatch and weave. Cold water forestalls fermentation and makes a viscous suspension, yielding the finest papers. The foothills of Echizen might not be able to grow much food, but Kawakami Gozen brought a craft suited to the mountains. Even the trees here grow longer fibers than those of the warmer valleys, giving paper strength and luster. Echizen became the hub of papermaking in Japan. The town was the exclusive supplier for aristocrats, shogunates, and governments. From these vats of water and fiber, Japan crafted its written culture. Later, as trade with the West commenced, the paper traveled to Europe, where papermaking methods were one thousand years behind those from Asia. Rembrandt preferred Japanese paper, likely from Echizen, for his etchings.

A scoop into the bath with a fine screen captures the maze of cellulose, locking the tangle as the sheet flattens. Repeated dips laminate the paper's surface. Water's capillary bonds, the same bonds that hold water within living plant cells, suck and mat the plant strands. Crushed by a press, water seeps out of the paper. The sodden sheet tightens as it oozes and drips. Finally the departing water draws the cellulose close enough that plant atoms find one another and bind, atom to atom.

The *kami* is both water and absence of water. Though her physical presence has evaporated, she lingers in every electrochemical linkage within paper. Billions of *kami* live in the charged atomic bonds that hold paper in being. My taxi driver was right to be confused, living as he did in a town full of papermakers. Deity, *kami*, 神. Paper, *kami*, 紙. The same sound. On our tongue and ears, Kawakami Gozen reveals her Shinto being: every profane scrap of paper contains the same unseen energy that dwells in her sacred shrines.

Her upper, mountaintop shrine is a small dwelling, shrouded for most of the year by protective tarps. Only on festival days are the veils removed. A statue representing Kawakami Gozen is carried through town in a golden litter, visiting papermaking workshops in a celebration that lasts for days. The lower shrine is built where the village meets the forested mountain slope. A *honden*, the inner sanctum reserved only for the *kami*, sits behind an offertory hall and oratory. The shrine is enclosed in a mossy, walled courtyard. A lantern-edged paved terrace leads to the *torii* gate. Around the shrine old *Cryptomeria* trees, cedars as tall as a rain-forest ceibo, recall the lances and straight-boled vigor of the Otaki warrior monks who lived here in the centuries before Kawakami Gozen arrived.

The lower shrine is cased in wood carvings of nesting birds, dragons, leaves, flowers, and acorns, a forest in which my eye wanders for hours. The shrine's three roofs, rising from the offering box to the recesses of the *honden*, are waves of shaped plank and bark shingle. This sylvan craft rests on a paradox: the shrine is held upright by the very molecule that papermakers must destroy. Without lignin—a stiff molecule whose struts and rings give wood its strength—twigs would be mere cottony threads and trunks could bear no weight. But lignin repels water and gums the molecular weave from which paper is made. Traditional craftspeople use wood ash and caustic soda to leach lignin from wet paper pulp. In modern paper mills, pungent sulfur does the same work. Worldwide, wood goes through its ablutions, purified before the *kami* can enter.

In carved wood the inner qualities of trees meet the outer design of artists. In paper too there is a meeting, but it takes place among molecules. The papermaker's artistry is to understand and work with fiber and water. Both tree and artist leave subtle imprints. A few papers contain flourishes—embedded leaves, fibrous strands, watermarks—that bring their inner nature to the surface, but most reveal themselves only under the fingertip and on the ear:

> Forged banknotes have a different timbre from real money. Criminals seldom find the right mix of plant species and water. The banker and printer hear the age and provenance of money in crack or slur, their fingers massaging sound from the paper. A connoisseur of cash hears paper's origin.

> I hold cotton stationery paper to my ear and caress the surface. Like a downy leaf, I hear softness on the stiff lamina, like a rake stroked over fine sand. When my fingers quicken: a metal skate on ice.

> Ganpi, a Japanese paper from the fibers of wild *Wikstroemia* bushes, is the "noble" paper, handmade and reserved for the finest lithography or the most expensive window screens. It barely sighs as my fingertips pass over its polished bone. What little sound emerges is high and even.

In two different pages of paper mulberry, kozo, I hear the divergent aftertones of the preparation process. The first page was made from unbeaten fibers. At my touch the paper crackles and swishes. Veins of hundreds of curled white fibers run down the sheet, so the texture of sound depends on the angle of my fingers. The second page was made by fibers that had been well beaten. A fine, strong paper that hums and purrs under my touch, a hint of abrasion, like the grit of fine powder.

Tissue paper. A quiet tear. Its fibers are flat and few, a bond easily loosened.

Newsprint pucks and pops on a dry day, gives an exhausted crumple when the weave is slackened by penetrating humid air, and falls silent after a week sitting in subtropical haze.

The drill sergeant of paper emerges with a rhythmic, gun-cocking sound from copiers and desktop printers. The paper shouts when disturbed and is strong in all dimensions. A uniform coating stabilizes the cellulose within.

I returned to the train station on foot, through streets empty of people. A few old men hung daikon radishes on racks behind their houses. One man passed with a wheelbarrow of kozo bark. Only when the hills had turned hazy with distance did the traffic swell and my feet carry me past shopping centers and expressways. In the nineteenth century Japan was home to nearly seventy thousand papermaking workshops. Now fewer than three hundred remain. Even as we use more paper than ever before—about 400 million tons annually, more than double our hunger of the 1980s—the page fades from our conscious awareness. In the industrial age, the sheet is almost entirely invisible, subservient to the inked messages on its surfaces.

Yet heedlessness is not universal.

Artists, printers, papermakers hear the *kami*. Some of this paper spirit trickles out on ritual occasions: wedding invitations, memorial books, birth announcements. In these noted missives, we hear and feel paper's import.

Refugees from Sudan and Bosnia have told me how, in flight, they treasured their few sheets, rationing their use centimeter by centimeter, every written word a joy of expression but a bite from the store.

And if ever the energy-hungry mills and electronic screens should stumble, what will outlive the present age will be written on handmade mitsumata, ganpi, cotton, or kozo: Kawakami Gozen's water and fiber.

Part 2

Hazel

South Queensferry, Scotland
55°59'27.4" N, 3°25'09.3" W

The tree's remains are swathed in plastic and enclosed in a cardboard sarcophagus. The archival box is marked with sample and location codes. Inside, labeled sacks keep order: Charcoal, Bone, Nutshell. I lift Charcoal and unzip its closure. Dozens of palm-size transparent bags, arranged in order of sample number, fill the cavity of the larger sack. I slide out the bags, releasing a sticky crinkle, the plastified sound of a well-ordered collection of artifacts. Tens of thousands of tree fragments, sorted by hundreds of hours of labor, nest within a hierarchy of numbers and names.

From the profusion of bags, I pick Charcoal-302-130 and tease open the airtight seal. Inclining the bag toward a glass dish, I tumble the nugget of charcoal from its home. It lands with a *plik* and sits isolated on a stage under the beams of a microscope's twin lamps. To the naked eye the sample is an irregular cube, each side about as long as a human fingernail. Although this is an old piece of wood, its charcoal darkness is as fresh as the burned remains of a campfire

extinguished minutes ago. With my eye to the microscope, the un-
differentiated char transforms into a landscape of cliffs sliced vertically by regular fractures. These fissures are the remains of rings of
cells within the wood. The fire burned away thin-walled wood, leaving blackened laminations. The fractures sink into the bulk of the
charcoal in curved lines, the tightness of the arcs revealing that this
was a small branch. Under magnification, and with bright spotlights,
the charcoal is speckled with silver, reflections from smooth surfaces
scattered over a background of dark pumice.

I do not break or slice the charcoal, as would be required for a
thorough laboratory analysis. Yet even without paring away sections
to mount on slides, I can see some of the identifying marks of the
wood. Growth rings undulate slightly. Pores are evenly diffused
through the wood, with no sharp difference within each ring between cells made in spring and summer. Rays, the spokes of tissue
that run perpendicular to growth rings, are thick, the result of individual rays aggregating into clustered lines. These marks are the
woody fingerprints of European hazel, as diagnostic as any test of
DNA or examination of leaf anatomy. Finer analysis by the archaeologists who dug these samples from the remains of a fire pit revealed yet more distinguishing features, all indicative of hazel: the
number of rays within each aggregation, the elongate perforations at
the end of water-conducting cells in the wood, and the five to ten
struts that cross these perforations. Some of the charcoal retained
the marks of fungal strands, showing that the wood had started to
rot before it burned. Every one of the thousands of pieces of charcoal
in the archival box bears the same signature. Whoever built and lit
the fire that produced this charcoal used nothing but hazel.

Charcoal comes in bags containing singletons or collections of
half a dozen fragments. Nutshells are bagged by the hundreds. The
sound as I open and tip Nutshell-302-231 is a tinkling sluice, like
pouring a jug of pennies onto a table. The nuts' identity is obvious,

even without a microscope. European hazel again. An apex marked with a low point, a globular shell with a flattened base, and slick shell walls. Where the walls have burned, a scallop shell of ridges runs from top to bottom. Even the nut's fawn color persists, although it is now charred. The sample contains no whole shells. Every nut is smashed into quarters or eighths. Someone had thoroughly processed these hazelnuts before letting them fall to the ground.

Because trees assemble themselves year by year from carbon dioxide gas, twigs and nutshells retain the signature of the year in which they grew, a signature written in the form of carbon atoms. The gradual, predictable decline of radioactive carbon-14 within the nutshell provides a clock. At first nutshells, like all living creatures, have as much carbon-14 as the atmosphere in which they grew. Then, as the radioactive carbon transmutes into nitrogen, the amount of carbon-14 declines, like sand running out of an hourglass. After about fifty thousand years, all the original carbon-14 has disappeared and the hourglass is empty. Up to that time, carbon dating is the best way to query the dead about their age.

We can make the carbon-14 hourglass more precise by calibrating it with the growth rings of trees of known ages. The widths of tree rings record the cycles of good and bad years, generally wide in moist years and narrow in drought. Piecing together the stories told by these rings can tell us, to the year, both how the climate has varied and how the carbon-14 signature of the atmosphere has fluctuated. Examination of old wood buried in swamps allows botanists to construct tree-ring records that date back for tens of thousands of years. Carbon-14 counts from nuclear physicists, combined with microscopy of dead wood, provide a chronometer of great precision. And so nutshell from the bags that I was holding found its way into a laboratory at the University of Oxford. There the carbon atoms were bombarded by metallic cesium ions, then accelerated through a chamber energized by 2.5 million volts. This electrostatic violence focused the vaporized nutshell into a beam that hurled itself into a sensor. The carbon atoms' answer, calibrated by tree rings: 8354 BCE, or 10,369 years old, with a seventy-eight-year margin of error.

Were it not for traffic congestion and aging bridges, the nutshells and twigs that I examined in the Edinburgh offices of Headland Archaeology would have remained buried under a suburban dog-walking park in South Queensferry, Scotland. The town's name hints at the area's problematic travel logistics. Just to the north, passage is blocked by the Firth of Forth, a wide estuary that slices across southern Scotland. Queen Margaret started a ferry service for pilgrims traveling to northern abbeys, locating the crossing at the narrowest point along the estuary. Ferries have connected North and South Queensferry for most of the thousand years since the queen's time. Then the Firth of Forth railway bridge opened in 1890, followed by a road bridge in 1964. Today no ferries run and the road bridge carries far more traffic than was predicted, on a structure that is inadequate for future loads. A new bridge is needed. Foundations must be dug.

The Queensferry Crossing road bridge, twenty-first-century Scotland's largest infrastructure project, sent bulldozers and archaeologists into suburban dog-walking fields. In Scotland almost every road-building project unearths some remnant of early agricultural settlements, medieval towns, or Victorian industrial development. Plans for road building therefore include time for archaeological surveys, to the frustration of motorists and the delight of those who wish to explore and learn from the past. Queensferry was a productive dig. Once the topsoil was sliced away, the sloping banks of the Firth of Forth yielded the remains of the oldest known human structures in Scotland, built by Mesolithic people on the heels of the retreating Ice Age.

Ten thousand years later, the glaciers are long gone, but the Ice Age feels close. On a sunny summer's day, the wind thrashes trees and raises whitecaps on the Firth of Forth. Rhythmic pounding comes from the construction site as tower footings are sunk below the water. Eider ducks, grunting as they beak the shoreline, shelter in the outlet of the stream that runs from the field. On the upper slopes, where the Mesolithic settlers lived, the wind shrieks and whirs in my ears. A meadow pipit's song cuts through with high, repetitive needling. Despite wind gusts strong enough to push me off balance, the bird flies a steady arc ten meters over the field as it sings. In winter sleety winds from the North Sea drill the cold through my hat and jacket. Geese and ducks power through the gales, flying upstream along the Forth River. These are birds of the north. Pipit, goose, and eider would have lived here ten thousand years ago. But today's climate is warm compared with that of the Mesolithic. Northern Scandinavia is a closer modern analog to the Mesolithic climate of Scotland. No wonder, then, that the first humans to colonize the site built a complex of sturdy shelters and wind-breaking berms.

Nine postholes large enough for heavy logs, angled inward, define the oval wall of the largest structure. Nothing remains of the walls, but deposits of clay hint that people may have used daub to seal cracks in twig-woven wattle. The post ring encloses a space of twenty-one square meters, the size of a modest room in a modern house. The floor lies knee-deep below the original land's surface, sunk perhaps for insulation and wind protection. One side of the interior is cobbled with river stones and a hearth sits in front of the cobble. At the center of the structure, an inner ring of smaller posts suggests the presence of a screen or frame. Pits for fire sweepings are dotted throughout. Rot has taken all that it can. The only identifiable remains are earthen holes, stone tools, and charred biological material. Samples Charcoal-302-130 and Nutshell-302-231 came from within this structure, sieved by archaeologists from a pile of black, silty sand. The samples came to us across the millennia because fire turned them to almost pure carbon. Such remains are nearly indigestible to the microbes that consume all else.

My first imaginings were of the hardness of life here. No doubt much of that life was difficult, but this site was also one of abundance. The estuary was close at hand. Fish and bird bones in the waste piles attest to food that came from the shores. Remnants of marine food are mixed with mammal bones of uncertain identity; there was likely no shortage of meat. One species, though, gave all the warmth and much of the food, and served as the keystone of these hunter-gatherers' lives: European hazel. Hazel branches were firewood. Hazelnuts were staple meals. In burning wood and roasting nuts, the people in this community inadvertently carried a few fragments of wood across many years, depositing them into archaeological bags.

Unlike the varied deciduous and pine forests of modern Scotland, vegetation ten thousand years ago was dominated by hazel scrub and woodlands. Some birch, elm, and willow mixed themselves into the forest, but hazel reigned, perhaps because of its tolerance of cold, damp weather and its ability to resprout after harvesting by humans. The remnant firewood from all the structures on the site is hazel, mostly small branches and twigs. Hazel is a relatively dense wood and therefore burns hotter than willow and longer than birch. Of the common tree species at the site, hazel was the best for firewood, not as good as the oak and ash that would, hundreds of years later, colonize the region, but the best at hand. The woodlands were therefore an excellent source of fuel, giving these Mesolithic settlers a local supply of all they needed for cooking and heating. Hazel resprouts rapidly after cutting, yielding within a year or two a new crop of burnable branches. The ubiquity and abundance of hazel wood in Scottish Mesolithic sites has led archaeologists to suggest that hazel woodlands were managed by coppicing, a deliberate strategy of repeated cutting aimed at producing large quantities of high-quality firewood.

Hazelnuts are so abundant in the structures that every step may have been a crunch of shell gravel. Hazelnuts contain a distillation of the best that trees can offer to the tiny clusters of embryonic cells buried at the centers of the nuts. The nuts are packed with all the nutrients that a newly germinated hazel needs: proteins, fats, carbohydrates, and vitamins. Sixty percent of the nut is fat and the remainder is protein and carbohydrate with little fiber. For humans two or three handfuls of nuts provide all the food needed for a morning's work. Hazelnuts keep well and can be stockpiled as insurance against lean times. Roasted, their shelf life extends into months with little reduction of nutritional content. Roasting also releases more flavor from the nuts. Regrettably, whether or how hazel was combined with other foods in Mesolithic cuisine is a mystery. The archaeological record, mostly derived from jumbled discard piles, has yet to reveal the particularities of individual meals.

In this Mesolithic village, as in many others across Britain, Scandinavia, and northern continental Europe, the hazelnut was the staple on which people built their lives. Archaeologists sometimes call this period in human history the "nut age." Later, when temperatures warmed and larger trees arrived, hazels declined, tightening the food supply. It is possible that the hard work of the Neolithic—tilling land to grow annual grains—was forced upon people by the decline of their favorite wood and nut crop.

Mesolithic hearths likely did more than cook, warm, and feed. They opened people to connections with others, deepening the human social network. Studies of extant hunter-gatherer cultures show that campfires change the nature of human conversation. During the day talk is of economic matters, complaints, or jokes. Around the fire the imagination opens and stories emerge. People talk of connections and rifts within human social networks, of the spirit world, and of marriage and kinship. Fire seems to anneal the human community, joining strands. Our minds seem particularly attuned to the sounds of fire. In the psychology laboratory, the blood pressure of experimental subjects drops and their sociability increases when the crackle of burning wood fills their ears. The sight of a soundless fire has little effect.

As with the wheat and oat seeds that would come with the Neolithic agricultural revolution thousands of years later, the Queensferry Mesolithic people fed themselves by tapping the cross-generational bequests of plants. In feeding on hazelnuts, these human colonists of a post–Ice Age land joined other vertebrate animals, especially jays and rodents. These animals, people included, were not merely passive followers of the trees. Their work as dispersers, like chickadees in a balsam fir forest or robins in a thicket of sabal palm, carried trees across the landscape. It is therefore impossible to draw a boundary between the fates of animal and plant species in these forests. Without animals most tree species would still be stranded along the Mediterranean coast, where they lived out the Ice Age. Without trees there would have been many fewer jays, rodents, and people on the postglacial landscape.

Hazel's particularly close bonds with birds and mammals allowed it to recolonize the land much faster than other tree species. About ten thousand years ago, hazel was moving north from the Mediterranean at about 1.5 kilometers each year, three times faster than oak. This rapid spread, combined with a relatively cool, wet climate, meant that for hundreds of years, sometimes thousands, hazel was the dominant tree species in northern Europe. In colder, wetter areas such as western Scotland, hazel retains its dominance, as if the Ice Age were lingering. A fondness for hazelnuts among birds and mammals fueled hazel's rapid spread. It is also possible, but the idea is controversial, that people may have accelerated the northward spread of hazel by deliberately carrying nuts over long distances to seed new land with orchards. In Scotland remains of wood, pollen, and human structures show that people and hazel arrived in the region at about the same time. If so, northern Europe's forests have lived in relationship with people from their origins in glacial rumble through to the present day. At no point were these forests primeval, unpeopled wilderness. Modern forestry in the region is therefore a continuation of an interaction as old as the forest itself.

As it spread across the continent, hazel also received help from belowground. The species' tolerance for a wide variety of conditions, including wet and cold soils, is partly born from a union of hazel roots with fungi. After chemical signaling among dozens of genes in both species, a sheath of fungus wraps the tree roots. Like the similar sheath in balsam fir, this cocoon acts as a mediator between root and soil, shielding plant cells from pathogens and feeding the roots with minerals. The tree responds with sugar from its leaves. Hazel ecology is a community matter, an intersection of plant and fungus, a fact not lost on truffle farmers, for whom hazel is a preferred fungus nurse.

The northern forests were born from mergers among ancestors of many species. The movement of trees across the landscape emerged from networked communities, with humans close to the center of the network. When Robert Burns crowned his spirit of Peace with hazel and William Wordsworth took his "nutting-crook" to bend hazel boughs, the poets owed their wreaths and meals to the relationships among ten-thousand-year-old forebears: humans, birds, fungi, and trees.

The ancient fire pits of South Queensferry sit almost directly across the Firth of Forth from another hearth. The modern incarnation of the hazel that fueled Mesolithic culture is coal—buried, transmuted wood, hundreds of millions of years old—inside Longannet Power Station. The burners inside Longannet yearly consume 4.5 million metric tons of coal. When it was built in the 1960s, Longannet was the largest coal-fired power station in Europe. The fire pit has grown in size since the Mesolithic, but the principle is the same.

Human society is, in this region as in many others, powered by fire from trees. By volume, coal contains five times more burnable heat than well-dried firewood, a convenience for householders and a boon to industry. Globally we burn 8 billion tons of compressed Paleozoic wood every year.

Longannet's location is no coincidence. Coal seams, including some exposed on the surface, thread the region's many geologic folds and cracks. Longannet Power Station is built on a deep history of regional dependence on coal, dating to the thirteenth century, when abbeys opened the first seams. The power station sits adjacent to some of these first excavations, connected to its own mine, which conveys coal directly to the burners.

Scotland's modern dependence on coal is partly a result of deforestation. By the fifteenth and sixteenth centuries, up to 95 percent of Scottish forestland had been cleared. The only fuel remaining was belowground. But even as wood disappeared from factory boilers and the hearths of most homes, trees literally supported the coal industry's work. Every mine shaft is a bet against the force of gravity acting on the weight of stone. Colliery memorial books and mine inspectors' reports are filled with the thousands of names of those whose lives paid for the mine owners' geologic gambling debts, pages of "killed by roof fall . . . coal fall . . . stone fall . . . fall of debris." "Pitprops," wooden staves and beams, were for centuries all that forestalled disaster. With little local supply, the docks along the Forth received shiploads of precut props from Russia, Scandinavia, and southern Europe. As late as the 1930s, the Scottish landowning gentry were exhorted to turn land into tree plantations to supply pitprops. For miners, keen attention to the sound of props was a method of survival. Before wood fails, it groans and pops. The cries of distressed wood were the signal to run before the roof collapsed. When mine owners switched to metal props, a seeming improvement given metal's superior strength, the warning fell silent. Metal fails catastrophically, with no preliminary call. In response, miners brought wooden props and wedged them against the metal, restoring the acoustic early-warning system. Trees, more than canaries, should be celebrated for saving miners' lives. Canaries were a twentieth-century innovation, used mostly by rescue teams after disaster struck.

Today the few remaining Scottish coal mines have no wooden pit-props. In belowground mines the roof is held by hydraulic posts connected to electronic sensors. Other mines are open pit, bringing the land's surface down to the coal, obviating the need for tunnels. Mostly, though, mines are closed or closing. Longannet Power Station will shutter within the year. The coal has not run out; enough remains to power Scotland for decades, if not centuries. A combination of new fossil fuels and concern for the side effects of burning coal combine to bring an end, for now, to hundreds of years of coal mining.

Imported natural gas is a strong price competitor for electricity production, abetted by a British tariff structure that penalizes Scottish power producers. Coal's downwind effects add to these troubles. One fifth of Scotland's human-caused greenhouse gas emissions stream from Longannet's single chimney, a smokestack nearly two hundred meters tall. Sulfurous and nitrogenous gases, combined with particulate pollution, also flow from the high orifice, despite new technologies to clean the exhaust. The principle of Longannet may be the same as the Mesolithic hearth, burning wood remnants to support our lives, but the principle's application yields problems.

Queensferry's hazel and Longannet's coal are two parts of a triumvirate of fire, all within sight of one another. The third member of the trio sits at the north end of the Forth Bridge, part of a proposed collection of power stations that would tap the fast carbon cycle of wood but do so on the scale of Longannet. If constructed, the Firth of Forth ports of Rosyth and Grangemouth would house Mesolithic-like wood fires, supplied and operated by twenty-first-century technologies. Once built, the facilities would join a growing national collection of power stations using pelleted wood to generate both piped heat for nearby industry and electricity for the grid.

The wood-pellet power plants replace coal with a "renewable" fuel, one that may contribute fewer greenhouse gases to the atmosphere. In 2009 the Scottish government, through an act of Parliament, set a goal of reducing emissions of greenhouse gases by 42 percent and 80 percent by 2020 and 2050, respectively. By 2013 Scotland's electricity was nearly half weaned from fossil fuels, with 44 percent of the supply coming from wind and hydroelectricity. On the hills and mountains overlooking the Firth of Forth, turbines screw Scotland's wind into the country's power lines. Wood-fired power stations will be, by some accounts, a necessary addition to complete the transition to renewable energy.

Coal is a much cheaper fuel than imported wood, so none of the pellet-burning projects would proceed without subsidies. On the face of it, subsidizing wood should loosen our dependence on ancient carbon and yield unambiguous benefits for the atmosphere. But, as with coal, what might seem a good idea in principle has hidden problems.

Hazel wood was not difficult to find ten thousand years ago. Gathering enough wood to power an industrialized country is a harder task, especially when the country has few forests. Other local users of wood, especially people in the furniture and wood-paneling industries, protest both the upward pressure on wood prices and government subsidies granted to their competitors in the wood-purchasing market. Local governments generally oppose any deforestation and deny building permits to smoke-generating wood-fired power stations. To pass muster with the Scottish government, proposals for wood-pellet supply must therefore procure their wood overseas, guaranteeing minimal effects on local forests. As with its former reliance on imported wood for pitprops in the mines, Scotland will not now supply its own timber for wood pellets. These Scottish decisions reflect a near-universal trend. As regions become more wealthy, they protect their own forests, increasing local forest cover. But demand for wood does not disappear. Rather, imports increase, resulting in deforestation in places distant from the leafy environs of the wood importers.

Even without local political pressure, the densely populated countries of northern Europe cannot meet their projected electricity needs with wood. They have too little land and forest. Wood pellets must therefore come from elsewhere. In the southeastern United States, where timber owners have few immediate prospects for vigorous sales at home, suppliers are happy to sign long-term contracts with European buyers. Peoples and trees that were loosely connected are now tightly yoked. Electricity switches in British homes fell the forests and tree plantations of the Carolinas and Georgia. Taxes paid in pounds flow to pellet mills built with dollars.

The docks of the southeastern United States, which formerly supplied slave-grown cotton and old-growth pine to Europe, are now fitted for wood-pellet exports. Wherever oceangoing ships can dock near supply routes from inland forests, stadium-like domes and warehouses are appearing. These giants are necessitated by the instability of their contents. Pellets are made by grinding, drying, and compressing wood. They are dusty and, if dampened, will decompose and heat up like a pile of lawn clippings or a barn of uncured hay. Heat added to piles of compressed energy and flammable dust risks spontaneous fire and explosion.

The environmental benefits of the transatlantic movement of wood pellets are controversial. Industry promoters emphasize the benefits of using waste wood to generate fuel to replace coal. Detractors fear the effects of increased timber harvesting on the biological diversity of southeastern forests and question whether transporting wood across the globe yields any benefits for the climate. Both sides have support from the accountants of greenhouse gas emissions. If pellets are made from sawmill waste or from thinnings of dense tree plantations, pellet fuel far surpasses coal in its carbon efficiency. On the other hand, if wood pellets are made from mature trees, carbon benefits decrease. If the trees come from native forests, not plantations, then pellets release more carbon to the atmosphere than coal. The effects of coal and wood pellets on biodiversity are likewise complex. Both forests and plantations can yield wood pellets and simultaneously provide habitat for many native species. Further, the economic value of wood products gives landowners an incentive to keep growing trees rather than convert the land to agriculture or housing. In using the forest, we protect the future of forest-dwelling biodiversity, merging our needs with those of other species. Yet if forest management results in large-scale conversion of native forests to plantations, as it has in some parts of the Southeast, many species of the native forest decline or are lost. We must then weigh the effects of these changes against the polluting effects of burning coal. In addition to coal's effects on global warming, mercury and acid in coal smoke degrade soil, harm trees, and poison waterways.

There are therefore no straightforward and universally applicable answers to the question of whether burning wood pellets or coal causes more harm to life's networks. The particularities of each forest determine the ecological effects of our actions.

Over the years, the consequences of our hearths have become harder to see and understand. A glance at hazel woodlands revealed the state of the energy supply for Mesolithic peoples. Their fires created smoky shelters, but the winds carried away these meager plumes. Coal's abundance and its downwind effects were evident in the centuries when nearly every Scottish house and lung was blackened. Thomas

Carlyle noted that, ever since the fifteenth century, the burning of a "certain sort of black stones" created a cloud over Edinburgh, giving the city a Scots name that it carries to this day, Auld Reekie, Old Smoky. Robert Louis Stevenson described the city as smoking "like a kiln," and Walter Scott, whose stone memorial in Edinburgh's city center even now retains the sooty stain, wrote that at twenty miles' distance, the smoke hovered as a "gosshawk hangs over a plump of young wild-ducks." In the twentieth century taller smokestacks and more efficient industrial burners tidied most of the soot while puffing world-changing carbon dioxide upward. Fuel and pollution became, for many, invisible. Even Longannet, though, is visible to the people of Scotland, as are its mines and ashpits. Wood pellets come from over the horizon, their origins unseen.

Global trade may allow the market to find each region's comparative advantage—the southeastern United States can grow more wood than can Scotland—but it forces energy users and policy makers to relate to benefits and costs of trade as abstractions. Ideas and statutes that live only in disembodied intellect are fragile, easily manipulated by both sides in a debate. This is as true for European "sustainability" regulations as it is for Amazonian *súmac káusai* removed from its forest home. Knowledge gained through extended, bodily relationship within the forest, including the forest's human communities, is more robust.

Imported fuels, pellets, or oil, renewable or otherwise, cut all strands of sensory connection to the sources of a society's energy. Our fuel tanks are networked to the world, but our minds and bodies are not. We are as dependent on fire as were the people of the Mesolithic, but we now stand at a great distance from the hearth. This, more than the details of policy regulations, is the weakness of globalized energy trade. Regulations can be rewritten; dislocation is harder to mend.

Europe's "green" is in fact a rainbow, many hues of sunshine from across the world, gathered by plants, then refracted through a haze of policy. This rainbow touches ground: wood, ethanol, and biodiesel come to European markets from the prairies and forests of the United States, Canada, Brazil, Argentina, Ukraine, Indonesia, and Malaysia. Extended experience of life in these places would add embodied knowledge to the renewable-energy policies of Edinburgh, London, and Brussels. There are truths that cannot be accessed through intellect alone, especially intellect that is not aware of local ecological variations. A few years of overseas collaborative work and listening would reconnect decision makers to people and places. In a world dominated by the arguments of white papers and scientific summaries, such a reincarnation of policy making would set a radical—literally "from the root," *radix*—example to other governments.

On my last visit to the Queensferry Crossing, the landscaping crew had been at work. Behind the chain-link wall, enclosed in rabbit-proof plastic sleeves, were knee-high saplings, among them European hazel, its ragged-edged round leaves poking from protective tubes. Like all parts of this construction project, the hazel arrived here via written specifications. Whether for concrete mixes, safety regulations, or tree plantings, engineers and planners work from checklists and percentages. Hazel is enumerated in HW1 and HW2, Hedgerow Plantings, at 14.0 percent, but is absent from MW1–4, Mixed Woodland. Whether or not Mesolithic people planted hazel as they colonized the land, we moderns have taken the task in hand to the nearest decimal place. With our help, the plants of the Mesolithic live on. As they inhale and grow, the atomic signatures of the twigs and nuts will bear the mark of Scottish coal and American forest, a jumble of ages and places.

Next to the hazel, on the bridges, tens of thousands of cars daily retrace the pious queen's crossing of the Forth. Under their wheels runs older traffic: Victorian iron, medieval coal ash, Mesolithic homes, and the leavings of Paleozoic forests. The passing vehicles are loud; they move fast. The sweep of new steel, arching nearly two hundred meters over the water, commands the attention of drivers and passengers, as should so fine a sculpture. On the road verge, young trees get barely a glance as they gutter in the ceaseless wind.

Redwood and Ponderosa Pine

Florissant, Colorado
38°55'06.7" N, 105°17'10.1" W

I wake to a flurry of scratching. A Williamson's sapsucker scuffles on the trunk of the ponderosa pine under which I dozed. The bird climbs with a steady rhythm, one bounce per second, bracing against the bark with a stiff, pointed tail. The scaly feet vault a few centimeters with each hop. The bird's head twists from side to side, glancing its beak over the bark's surface, tongue-jabbing ants. Williamson's sapsuckers raise their young almost entirely on ants, so this bird likely has a squalling brood of mouths waiting for it in a tree hole nearby.

Tail, feet, and beak chafe and grate on the bark, a bustle of sound as the bird works. The noisiness of this individual is no exception. These sapsuckers scrape and thud as they take their business through the forest. I found another bird yesterday by following the commotion created by its reworking of wells and wounds in a Douglas fir's bark. The chiseling blows were audible long before I could see the bird supping at the tree's bleeding sap. This coniferous ooze is the favorite food of adults, providing the sugars needed for the birds to fatten themselves for breeding and to squeeze through winter's lean times.

The bird's passage above my head has flaked the bark, releasing wafts of scent from the tree's sun-warmed, friable surface. The golden sap between dark plates of ponderosa bark has the vigorous odor of rosin and turpentine: oily, acidic, and bright. But unlike the aggressive, spiky odor of other pines, ponderosa's aroma has smooth, sweet edges. A hint of vanilla or buttery sugar mingles in the resin. Attentive human noses, and perhaps the tongues of woodpeckers, know that the nasal tone of ponderosa varies geographically, faint in the Northern Rocky Mountains, stronger, with a twist of lemon rind, along the Pacific coast. The scent is a deterrent against attacking insects. Sticky resin gums and traps wood-boring insects, and resinous chemicals are poisonous in large doses.

Resinous defenses are adequate against most insects and in most years, but lately ponderosa and many other pines have been dying by the millions, killed by beetles. Paradoxically, the odors that protect the tree are the same ones that lead these beetles to their target. What is protected is valuable; defense is also advertisement. The pine beetles sniff the air and fly upwind to the pine. Once arrived, they bore under the bark, feed on the tree's living tissues, and, if the beetles are numerous, kill the tree. These attacks have become so widespread in the Rocky Mountains that it is common to see whole valleys turn from living green to dead-needle brown and finally to bleached-wood gray.

Pine beetles have always lived in these mountains. But now their populations are surging, pushed upward by a landscape full of drought- and heat-weakened trees. Whether the sapsucker will be here in a few decades no one can say, but some projections show the species on the road to extinction. The bird's fate depends on how ponderosa and other trees fare as the elements—wind, water, earth, and fire—are reborn in a changing climate.

I sit up in my fragrant bed of needles and resume my days-long vigil. I'm in a copse of ponderosa at the edge of an alpine meadow in the Colorado Rocky Mountains. To my left an open stretch of grass and herb crosses a shallow valley, then rises to meet more pines on the ridges half an hour's walk from here. To my right is a crumbling slope of mudstone and shale, partly scraped away to reveal the Big Stump, the base of an ancient redwood trunk, one of two dozen large fossilized redwood stumps along the trails of the Florissant Fossil Beds National Monument. The monument was created to protect and celebrate these fossils of petrified wood, but it is often the modern creatures that first catch our attention: bobcats sleeping amid wildflowers, ravens and hawks calling as they chase one another, grasshoppers clattering ahead of us on the trail through the pines.

"What's that *huge* sound?" calls a young girl to her family as she ambles in pink trousers toward the ponderosa copse. She is an attentive child. Of all the visitors I've encountered here, she's the only one to remark on the trees' song. She's right: *huge* it is.

A glance of wind sets the pines huffing. A modest breeze evokes an urgent hiss, steam escaping from a dangerously pressurized valve. A gust is like a landslide, sand avalanching down a gully; a sound of this kind in the maple and oak forests of my home in the eastern United States would send me scurrying for cover, an eye on the canopy for snapping trunks and falling limbs. But here the pines carry no such warning in their shouts.

The hugeness in our ears has its origins in the ponderosa's stiff needles. The leaves of other trees accommodate themselves to flowing air, but ponderosa needles do not flex. Branches and twigs bob in the wind, but needles are unmoved. They harrow the wind, fracturing it with thousands of unyielding tines, scoring the air with violent grooves. There is no resonance in this score, no lingering energy from flapping and trembling leaves. Instead, the tree reports the wind's character second by second, whipping to a higher register as a gust passes, then tapering, bulging, or dying away in the air's changing movement.

John Muir also noted the ponderosa's sound and his description puzzled me. In the trees' reaction to the wind he heard the "finest music" and a "free, wing-like hum" from the needles. Where was the harrow, the urgent distress? Muir heard Aeolian harmonies in his piney mountains, but I heard Ariel wailing from his prison, flailing the sky in torment. These divergent experiences might reflect a difference of disposition; Muir's relentless ecstasies are hard to match. But the writings of plant taxonomists later taught me that Muir and I were hearing different dialects. The ponderosa pine is a variable tree. Not only does the aroma of its resin differ from place to place, but the shape and stiffness of its tongue also comes in regional variants. The trees that I heard in the Rocky Mountains have needles that are only half as long as those in Muir's California. Thick-walled cells under the needles' skin add rigidity, making Rocky Mountain needles more like wire brushes than the mare's tails that grow along the Pacific. Shorter, stiffer needles make a fiercer sound. Ariel, it seems, is a happy captive in California, singing sweetly from trees rooted in relatively moist soils. Only in the dry Colorado mountains does he vent his groans from needles adapted to dry summers and heavy loads of winter snow.

Our unconscious triggers of fear must tune themselves to the sensory particularities of place. My conscious mind knew that no windstorm was threatening me; the memory of familiar trees from elsewhere told my body a different story. No doubt the attentive girl also lives among other trees. The incongruity of the ponderosa's noise struck her as strange. What is true for the sounds of the forest is true elsewhere: a country mouse cannot sleep in the city for sirens and shouts, yet the city mouse is set on edge by rural silence or a cacophony of late-summer katydids around a wooded cabin.

There are sounds in these trees whose pitch is too high for any human ear. These ultrasonic clicks and fizzes reveal the hidden drama inside trees' water conduits. Because availability of water is so often the determining factor in the vitality or decline of plants, eavesdropping on ultrasound takes us to a tree's heart through the sounds of water in its twigs and trunks.

Every tree leaf, including the needles of ponderosa pine, has dotted over its surface hundreds of pores, the stomates, through which gases pass into and out of the leaf. The pores are active, opening and closing through the action of two lip-shaped cells that purse or gape, like miniature mouths. When the lips part, they allow a rush of air to bathe the interior of the leaf, supplying carbon dioxide to the photosynthetic cells that feed the plant. Water vapor streams out of these parted lips, drying the leaf and dragging water from the roots. If the soil is wet, this presents no problem. But when the soil dries, roots cannot replenish the leaves. Mouths must then close to prevent catastrophic drying of the leaves' interiors. Lack of water therefore chokes the nourishing flow of air. No water, no photosynthesis.

I strap a thumb-size ultrasonic sensor to a ponderosa twig and wire the electronic device to a computer. Then I wait, "listening" through the intermediary of a graph on the screen. Every time the twig releases an ultrasonic pop, the graph's line jolts up by one step. Each individual pop reveals little, but over many hours, patterns emerge: vigorous ultrasonic activity when the twig dries, relative quiet when it is well supplied with water. The briskness of acoustic activity calls out, hour by hour, how the twig's water vessels are faring.

Sudden breakages in the columns of water that run from the trees' roots to their crowns cause these ultrasonic sounds. Water flows through interlocking hollow, wooden cells, each cell about as tall as an uppercase letter on this page and as narrow as the finest human hair. When the soil is moist, the water moves freely, tugged upward by evaporation from the stomatal mouths pulling on the cohesive columns of water. But when the roots can no longer replenish the movement of water and when the tug from dry winds gets too strong, the silk-thin water strands break. Immediately afterward, an air pocket explodes inside the cavitated cell. Like rubber bands stretched beyond their limits, these breakages snap. At the tiny scale of cells, the sounds are pitched so high that they soar over the upper limit of our hearing.

For the tree, ultrasonic snaps are the sounds of mounting distress. Air pockets block water's flow. These blockages can happen anywhere from root to needle. All trees experience these minute interruptions of the movement of water, but pines growing in dry soil are especially vulnerable. In some ponderosa trees, especially young ones, air pockets clog nearly three quarters of the roots by the end of summer. When moisture and cool days return in late autumn, many of these roots will recover, but this is of little help in summer, when a tree needs to feast on air and sunlight. Lack of water will therefore weaken or kill a tree by starvation. No nourishing carbon dioxide can flow through stomatal mouths that remain closed for want of water.

My electronic apparatus also detects the motions of even tinier air bubbles in the twig. The bubbles cluster along the edges of the water-conducting cells. Like a wall of balloons, the lining is elastic and it absorbs then releases pulses of pressure. When cells dry, then rehydrate, the bubbly lining shifts violently, releasing a crackle of ultrasound. Water pipes in trees are therefore like those in old houses, knocking and groaning as water moves, but with sounds transposed many octaves higher.

The forest sizzles, but our ears fail us. What might we learn with better ears? At the very least, through the trees' ever-changing squeals and snaps we'd be aware of the dynamism under the deceptively still plates of bark. Robert Frost lost "all measure of pace" amid the noise of trees. We and Frost are perhaps lucky. If we could hear the inner cries of every twig in the forest, ultrasound would truly unfix us.

My electronic gadgetry is a poor substitute for bodily awareness of the trees' sounds. Nonetheless, a story emerges through the scrolling graph on the screen. The tree is quiet through the morning, signaling an orderly and abundant flow of water from root to needle. If the previous afternoon brought rain, the quiet is prolonged. The tree itself makes this rainfall more likely. Resinous tree aromas drift to the sky, where each molecule of aroma serves as a focal point for the aggregation of water. Ponderosa, like balsam fir and ceibo, seeds clouds with its perfume, making rain a little more likely. But rainy afternoons are uncommon in late summer; a rainstorm that soaks one valley may bring not a drop to the rest of the mountain.

After a rainless day, the roots' morning beverage is brought by the soil community, a moistening without the help of rain. At night tree roots and soil fungi conspire to defy gravity and draw up water from the deeper layers of soil. The lacework of roots connected to threads of fungi acts like a vast piece of blotting paper, sliced and loosely woven into the soil's depths. This paper is made not of the disorganized cellulose found in paper but from cellulose sewn and spliced into tubes and cell walls, a ramifying network of slender conduits. Water is attracted to the slight electrical charge on the roots' cellulose molecules and in the cell walls of fungi, then easily slides along the tubing, flowing as the laws of physics dictate, from wet to dry. So even without the power of the Sun pulling water from the surface of the soil and from plant leaves, water moves upward through the night.

Evening's dust-dry soil surface is thus dampened by morning, a process that keeps many trees alive by delaying the blockages caused by air pockets in snapped water columns. It is not just trees that benefit from this nocturnal reverse rainfall. Grasses, herbs, microbes, and soil-dwelling animals like springtails, mites, and beetles also get a boost from the upward flow caused by the symbiosis between root and fungus. These effects on the wider community are poorly understood, but it is not too radical a speculation to say that without the union between plant and fungus, these alpine forests and meadows would dwindle, pinched off by the meager rains and dry wind.

By noon the graph tracking ultrasound inflects upward. The soil has dried and the twig's water columns are crackling. My own physiology echoes the tree. My lips crack and my water bottle has run dry. Mornings here are pleasant, then discomfort sets in as the long day's exposure to dry air and high-altitude sunshine takes its toll. I doze in the piney shade, but the tree has no such luxury. Now the ponderosa's mettle is tested. In the afternoon smelter, most tree species would burn away, turned to ecological dross by their ruined water conduits. The species that survive, the gold resting in this alpine crucible, are those whose physiology can withstand drought. Ponderosa's nighttime lift of water helps, but this is only one of the species' adaptations. As soon as roots dry, the needles' pores squeeze tight. This firm closure, combined with the needles' thick skin and waxy coat, clamp the flow of water with a vigor unmatched by species adapted to wetter soils. Like those of all pines, the ponderosa's water-conducting vessels are narrow and connected by closable perforations. Air pockets are therefore confined to single cells, a first aid mechanism lacking in many other tree species. Ponderosa can, when needed, be supremely miserly with water.

This thriftiness with water starts early. Ponderosa seeds germinate on the soil surface and need just a sprinkling of rain to break open and grow. Too much rain is harmful, as it encourages smothering growth of surrounding grasses. Once germinated, the tree lances the soil with a vertical shaft, a taproot. By the second year of the

tree's life, when the seedling is an ankle-high tuft, this root can penetrate the ground to a depth of half a meter, with lateral roots splay-
ing to the side. By adulthood, if no rocks or other trees have impeded
the roots' growth, the taproot can sink twelve meters down and the
laterals extend more than forty meters. These roots are wed to an
even more extensive net of soil fungi. The tree that we see above the
ground is the sun-gathering appendage of a community of roots and
fungi, a chimeric water-seeking subterranean giant.

Sometimes the smelter's heat bursts into flame, and here too ponderosa is prepared. Rain may be unreliable in late summer, but lightning is not. Passing electrical storms have curtailed several of my
visits. Many trees, including the one I sit under, have furrows sliced
into their bark, running from canopy to root. Bark bulges from the
sides of these gashes, the trees' attempts to seal the wounds, but
lightning scars are seldom fully healed and bare wood remains exposed in most, faded to gray by the alpine sun.

The dry needles and grass on which I sit make excellent kindling.
When lightning ignites the bed, flames crawl or race through the
forest understory. These ground fires leave a blackened forest, but
older ponderosa trees are unburned, protected by their dragon-skin
bark, a shield of thick plates that resist heat and flame. Aspen and
other tree species are less fire hardy, and a low-intensity burn clears
away ponderosa's competition. When fires sweep the ground every
decade or so, ponderosa fares well.

The dragon is not impervious to flame. Some fires burn hotter
and higher. They climb from the ground to the canopy, then feed on
the thick growth of branchlets and twigs. If more than half of a ponderosa's crown is burned, the tree will die. Fierce burns clear whole
mountainsides of living trees. The forest returns only through the
decades-long processes of seed germination and sapling growth.

The character of the forest therefore depends on how often the mountains are visited by fires of low and high intensity. The rhythm of these burns depends on many factors: foresters and landowners have extinguished smaller fires, building fuel for larger conflagrations; piney aromas of drought-weakened trees attract lethal swarms of tree-killing beetles, leaving behind forests stacked with dry firewood; and, most of all, variations in moisture and heat—short-term fluctuations in weather and longer trends in climate—fuel or suppress fire.

We can glimpse the effects of a changing climate in the soils downstream from mountain forests. When mountain watercourses reach the valley, they slow and meander. Any sediment in the water drops from the sluggish flow and builds alluvial fans, a geologic record of upstream erosion. Waterways in unburned forests run clear, carrying no debris to the valley. Low-intensity fires send downstream pulses of sediment and charcoal. After a severe burn, creeks are choked by jumbles of burned wood, boulders from landslides, and eroded soil of all kinds. By digging through these strata of downstream alluvium we can unlayer time and reconstruct fire's history.

The last eight thousand years—a period stretching from just after the last ice age—have burned with an unsteady flame. In some centuries fire merely sputters; in others it flares. The vicissitudes of climate seem to be the cause of this irregularity. During the Little Ice Age, from the fifteenth to the nineteenth centuries, cooler and wetter conditions prevailed. This damped big fires but also encouraged lush growth of grasses, fuel for low-intensity fires. The sediment from these times indicates that forests burned often, but the fires were feeble. In contrast, the earlier Medieval Climatic Anomaly, a warm period at the turn of the first millennium, brought decadal droughts to the West. In this parched time enormous, hot burns left thick deposits of washed-out mountain.

Given fire's great variability, pointing to a "normal" fire regime is impossible. Instead, careful study of soil and air can tell us fire's current mood and maybe let us glimpse the future.

In a mountain ravine a few kilometers downhill from the Florissant ponderosa pine sits the town of Manitou Springs. Yesterday a flash flood surged through town. Today I'm in a basement of one of the downtown stores, shoveling mud with other volunteers. A tendril of odor from a makeshift kitchen slides down to where we work, overpowering for a moment the stench of mud, rot, and ash. We're temporarily stunned, our work in this dark cellar halted by conflicting signals: sweet, homely perfume in a pit of devastation.

The storm that caused the flood was a modest one. A shot glass under the rain cloud would barely have filled. But the rain fell in one burst on more than seven thousand hectares of bare ground, the aftermath of a fire so big that it had gotten a name, the Waldo Canyon fire, and become a much-filmed celebrity on the television news. The fire had burned the previous summer, turning pines, aspen, and spruce to vapor and smoke. Hundreds of houses were torched.

What is left on the mountain is usually called a burn scar, a wound on the old forest, but the naked soil, mixed with charcoal, also presents the face of a newborn sylvan infant. Regardless of the names and emotions that we project onto the fire's remains, the mountains' surfaces are now at the mercy of the laws of physics. Friction among soil particles holds slopes in place, but only in dry weather. Light rain increases adhesion among the particles, giving inclines the temporary stability of sand castles on the beach; but heavy rain lubricates, weighs, and topples the castles, sluicing soil into stream channels. When the flood reached Manitou Springs, it was a wave of roaring, roiling darkness twelve feet high, coursing through a creek bed that normally runs in quiet splashes, ankle deep. One man died; many more were injured. Houses disappeared and stores had their inventory turned to sodden heaps. Future geologists will examine the downstream alluvial fan and conclude that this was a big fire.

The flood painted a uniform coat of mud to eye level on the walls of the basement in which we work. A few pine needles, aspen leaves, and snapped twigs decorate the top of the wall stain, pieces of streamside forest stranded after surfing the crest of the flood. The floor is smothered by ankle-high ashy ooze, a heavy slop spiced with shards of broken glass and splintered wood. Bucket by bucket we haul the debris. Metal shovels scrape on the concrete floor. Mud hits buckets with a punch. We grunt as we lift and carry. Some buckets go upstairs to bulldozers on the street; the rest of the mud is dumped out of the back door, dropping into the creek, which now runs lower, full of dark silt.

Thirty years ago, rarely more than eight thousand hectares of Colorado forest burned each year. Now many years top eighty thousand hectares, an area the size of a large rural county. The U.S. Forest Service now spends over half of its monies on fighting fires, up from one fifth of the budget ten years ago. Ash from ponderosa and other burned trees has a global reach. Soot from these fires, and from those in the Canadian boreal, reaches all the way to Greenland. There the ashy fall darkens ice sheets and causes them to absorb sunlight. The ice sheets melt as the Sun's warmth soaks in. Closer to home, ash and eroded debris clog Colorado reservoirs, increasing the cost of drinking water and slowing hydroelectric production.

Elsewhere fires are transforming tropical and boreal forests alike. In Southeast Asia forest fires kindled to clear land for agriculture are so large that smoke can envelop many nations. In these countries inhalation of particulate matter—tiny airborne flakes of burned rain forest—is now such a major public-health problem that multinational agreements attempt to control the burns. Drought-induced fires in the Amazon are unraveling the ecological integrity of the region, even in areas that are free from large-scale land clearing. In the far north the boreal forest is burning at such a rate that regrowth cannot keep pace. Summed across the world, all these fires are speeding the flow of carbon to the atmosphere. Forest fires are responsible for one fifth of the extra carbon dioxide released into the atmosphere since preindustrial times.

When geologists dug into alluvial fans in the western United States, they found the signature of the Medieval Climatic Anomaly, the most fiery time in the West since the last ice age. Our climate now seems to have surpassed those medieval extremes. The U.S. West is so dry that the western side of the North American continent has risen by several millimeters in the last decade, lifting as the watery load is taken from its back. Increasing temperatures in spring and summer, combined with early snowmelt, have lengthened the western fire season, and the area of land burned each year has increased by six times since the 1980s. Models of the next one hundred years all converge on the same prediction for the region: the ancient decadal droughts will look "quaint" compared with what is coming.

Wood and oxygen, the ingredients of forest fire, are born in the same place, in the process of photosynthesis, the legacy of the Gunflint bacteria. This unstable mixture of gas and botanical combustibles has been burning ever since plants colonized the land. Lately, for the last few hundred years, we've been in a relatively cool and wet time. So we built our infrastructure—town centers, reservoirs, suburbs—as if fire were a rare aberration. That time is over. We're all now downstream of the burn.

In Manitou Springs the dankness smothers the smells of cooking, removing us from the familiar, the odor of home, leaving us with only wet ash and mud. Our bodies return to work, shoveling sludge, then hefting buckets. Hour by hour, human muscles find the rhythm of their new world.

Thirty-four million years ago a series of volcanic eruptions in Colorado created flows of lava and debris that would, if they happened today, bury not just streamside basements but whole cities. One of the volcanoes, the Guffey, towered over the Florissant valley. Today its eroded remnants are just a few small ridges, dwarfed by the dome of Pikes Peak. In its time Guffey's moods, and those of its volcanic neighbors, dominated life in the region. The volcanoes periodically exploded and rained molten rock over their surroundings. At other times they sat quietly, burping lava and puffing ash. These geologic exhalations covered the landscape with ash, mud, and rock. Like the modern unstable mountainside above Manitou Springs, this accumulated debris would sometimes descend in smothering sheets, turned to slurry by water and snow. Some of these flows buried downslope valleys in several meters of ooze and fractured rock.

In one valley near Guffey, a debris flow five meters deep buried a forest of redwood trees. These were large trees, so most of their trunks extended above the flow. This standing wood soon rotted away as the suffocated tree roots died. Volcanic mud entombed the base of the redwood trunks. The mud was thick enough that the air's oxygen could not reach the buried stumps, choking the bacteria that would otherwise have degraded the dead redwoods. With biological decomposition vastly slowed, the long process of turning wood to rock began. Mineral-rich water seeped from the ashy mud, soaking every cell of the buried trees with dissolved silica. Gradually, over hundreds of thousands of years, the silica crystallized, replacing wood as it disintegrated. The growing crystals followed the shapes of plant cells, creating a rocky imprint of the tree. After millions of years of suffusion with mineral-rich water, the redwoods became petrified.

Today stone redwood stumps liberally stud the ponderosa forests and grassy meadows of Florissant Fossil Beds National Monument. Tree rings in the trunks, the twist of wood fibers, the shape and texture of root buttresses are all preserved. The fossils look papery and fragile, like friable old wood. But the sample of petrified wood in the visitors' center is shockingly unyielding, with the heft and hardness of iron. The mineralogical nuances of the wood stone are visible in its colors: dark streaks of manganese in the roots, iron oxides glowing orange in vertical faces, a flush of yellow where the iron has bled away. The minerals are overlaid with the lime green and russet of lichenous growth on the rock surface. As the angle and quality of light vary through the day and through the seasons, the hue and luminance of the colors change, animated by the touch of the Sun. In winter's oblique light, the rock is fiery coal; in summer the stump turns sulfurous with slashes of bleached white marble.

I sit with the ponderosa pine, next to the Big Stump. Before the volcanic flow, this redwood was seventy meters tall and more than seven hundred years old. Now it is a fragmented stone column three meters tall and ten meters around. The former landowners here scooped away the soil and rock around the stump, leaving the fossil exposed in a half bowl cut from the hillside.

For such a long-dead creature, the stump is an acoustically lively character. In the summer violet-green swallows wheel around the exposed trunk, chattering as they ambush flying insects. They land on the stump and the exposed mudstone to mouth at crawling insects and pick at soil fragments. Mountain bluebirds gather on the stump to feed their squalling youngsters, to purr at mates, and to snap their bills at rivals. Their toenails scrape at the rock as they land and shuffle. A hummingbird buzzes face-first against the stump, investigating a streak of flowerlike orange in the rock. Grasshoppers clatter in the soil at the stump's base and chipmunks climb vertical faces to survey their territories and stutter alarms to hawks and ravens overhead.

In autumn birds gather to feast on seed. Ponderosa cones open, grasses bend under the weight of their grainy heads, and herbs drop seeds to the ground. In this fruitful time the stump again makes a convenient hub for activity. Bluebirds meet and give one another trilling coos, and nuthatches try to pound pine seeds into the stumps' fissures before abandoning the effort and moving to the soft bark of a living pine tree. Purple finches and Steller's jays circle, calling, before flocking to the ponderosa cones. When the sun warms the ground, grasshoppers emerge by the dozen, pulsing dry, ratchety trills as they fly in the amphitheater around the stump.

Fewer animal sounds enliven winter's air. The wail of ponderosa needles dominates, interspersed with the *kok-kok* of passing ravens. Wind bends spent grass stems to the ground; as they move, their sharp tips etch curved lines on the snow's surface, the scratch of a pen on rough paper. Snow falls in clumps from pine needles, a hiss, then a muffled blow.

Sounds from people mingle unpredictably with the plant and animal sounds. Airplanes pass, smearing the air with growls. Hikers tromp, crunching petrified tree fragments underfoot. At the Big Stump amblers pause and comment on the fossilized tree's girth. Cameras shoot the *ack-ack* of their pixel flak at the stump, the ponderosa trees, the meadow, then the gunners stride away. A wood chipper growls and clanks at the visitor parking lot, the sound of land management by the Park Service. Hoping to forestall a fire, the rangers have thinned a piney hillside and hauled trunks and branches out of the forest. Wood and metal clash in the grinder, Ariel screams in the fast-striking mill wheels, and a spew of lacerated wood accumulates in a steaming pile, a modern Big Stump.

Impressive as they are, the redwoods are not the most exquisite or the most scientifically valuable fossils at the site. Volcanic flows not only buried trees but also dammed the Florissant valley, creating a small lake. A vigorous canoeist would take half a day to paddle the twenty-kilometer length of a modern equivalent. The original lake is now entirely dried, and from the remnants of its bed scientists have pulled some of the world's most beautifully preserved fossils. Guffey's intermittent activity rained layers of ash and clay into the lake. The thin strata of ash were stacked between dustings of microscopic skeletal remains of algae. These fine-grained, layered sediments trapped leaves, insects, and other living creatures, holding them like flowers between the pages of a book. As the sediments on the lake bottom accumulated, the ash and algal skeletons gradually turned into a rock known as paper shale. Now, with a gentle tap from a hammer, the thin sheets of this rock open, cracking the spine of the ancient book to reveal stony imprints of the flowers that fell between its pages.

The Yale Peabody Museum houses some of these fossils in its collections. To hold them in my hands is a marvel. Looking as fresh as leaves and insects newly fallen onto water, the fossils belie their age. Under a hand lens, every ramifying vein on a fern leaflet is visible. With a microscope I gaze at the fine details of floral anatomy, the shape of pollen grains, the tile-work arrangement of cells on a leaf's surface. Even with the naked eye, particularities startle. The irregular holes on tree leaves look as if the caterpillars that chewed them wandered away and will return by evening. Spider fangs, crane-fly antennae, ant eyes: the thin, delicate parts that are entirely missing from most fossils are present here.

Florissant paper shale is an Alexandrian library of paleontology, but created and preserved by volcanic fire. The tomes of shale are more than thirty million years older than the papyrus scrolls of the Mediterranean scholars, and the shale carries within it the stories of thousands of species. These stories combine to let us see and hear the ecology of Florissant in what is, in geologic terms, the recent past. Viewed from the present day, 99 percent of life's history had already unfolded by the time Guffey exploded and trapped the Big Stump in mud. The precise etching of ecological memory in Florissant's fossils tells us that the last 1 percent, although brief, has been a restless time.

The Big Stump and its companions reveal the central message of the Florissant fossils: This was a much warmer, wetter valley 34 million years ago. Modern redwoods grow in the Americas only along the temperate Pacific coast. In today's Colorado, summer's drought and winter's chill would snuff any redwood, seedling or giant. Along the shores of the ancient lake and its tributary streams, though, redwoods thrived. We can estimate the long-dead redwoods' growth rates from the growth rings preserved in petrified stumps. The wide rings of the fossil trees tell us that they outstripped their modern counterparts. Florissant's climate was even more balmy and moist than that experienced by today's redwoods.

Dozens of other plants grew with the redwoods. Oaks, hickories, and pines lived on the higher ridges, while poplars and ferns grew closer to water. A few of the plants have no clear modern relatives, but many are species from familiar genera. Grapes, greenbriers, blackberries, locusts, serviceberries, palms, and elms were all present. The modern flora at the site include almost none of these plants. By comparing the assemblage of plants found in the shale to the habitat preferences of modern descendants, botanists can reconstruct the climate of ancient Florissant. The conditions experienced by the forests in the temperate mountains of Southeast Asia or central Mexico are perhaps the closest modern analogs. Summers are hot and humid; winters mild, with rare freezes. The average annual temperature in ancient Florissant was at least ten degrees Celsius warmer than it is now. For comparison, policies addressing modern climate change have as their goal an increase in the global average temperature of no more than two degrees. We'd need to overshoot this goal four times before we'd reach Florissant's temperatures.

Animal fossils confirm the plants' testimony of a warm climate. Cicadas and crickets whined from the redwoods. Crab spiders lurked inside flowers and hundreds of species of beetle roamed the moist leaf litter. Fireflies lit a forest understory where orb-weaving spiders strung their silk. Water scorpions swam alongside bowfin fish in the

lake, and plovers walked the shores. Compared with modern dry
meadows and uniform stands of ponderosa, these ancient forests
were extraordinarily dense and diverse, their overflowing cup of life
supplied by regular, tepid rains. Periodically Guffey would annihi-
late part of the forest in a spasm of geologic ire, but the forest would
then return and leave yet more rocky volumes crammed with impressions of biology.

The remarkable preservation of Florissant's fossils, combined
with the site's high biological diversity, makes the shales and redwoods famous among paleontologists. But these fossils represent
more than the ancient life of one exceptional corner of Colorado.
Florissant's creatures were part of a great flourishing of life that happened during the Eocene period. The Eocene climate was unsteady,
but it was always much warmer than today's world. Concentrations
of carbon dioxide were at least double modern values, perhaps even
ten times as high. Releases of methane from undersea vents and fissures added more heat-trapping gases to the atmosphere. The planet
was not so much a greenhouse as a sauna. At the peak of the Eocene's
heat, the planet was hot from pole to pole. Lush forests grew in what
is now the treeless Arctic, in temperatures thirty degrees warmer
than modern Arctic conditions. Palms grew in a frost-free Antarctica. The prodigious quantities of Eocene atmospheric carbon dioxide appear to have come from many sources: volcanism, weathering
of carbonate rocks, releases of gases from the ocean and swamps,
and changes in the carbon stored and released by algae.

The hottest part of the Eocene came tens of millions of years before Guffey buried Florissant. By Guffey's time, at the very end of the
Eocene, carbon dioxide levels were falling and Antarctica had turned
from a warm, vegetated land to an ice sheet, putting a cooling snow-
cap on the world. Global temperatures were already much lower than
their Eocene highs. So although the fossils of Florissant were buried

in a warmer climate than that of modern Colorado, their era was cold compared with what had come before. The Florissant site's age places it on the hinge of what paleontologists call the transition from a "greenhouse" to an "icehouse" world. Because of the poor subsequent fossil record, no one can say when the redwood forests finally disappeared from Florissant, but they likely did not persist much beyond the end of the Eocene.

The deepening chill that had started when mud buried the redwoods continued to the modern day. Like the Eocene's climate, the temperature of the intervening years has staggered up and down. When examined from a distance, though, the drunken walk takes a downward path. We owe the origin of our own species to this cooling trend. When Africa's forests retreated in a particularly cold and dry spell, our prehuman ancestors strode into the emerging savannas and grasslands. *Homo sapiens* evolved from these apes of the open country, and all our species' history has unfolded in relatively cold times. The calm that I feel as I survey the scenery around the Big Stump—open vistas of grasslands and tree copses, created by cool aridity—is perhaps a judgment of the landscape wired deep in my human mind. The opossum-like creatures hunting insects in the branches of Florissant's Eocene redwoods or the twig-nibbling miniature horses below would likely prefer the view of a tall tangle of vegetation, seen through a haze of rain.

An affinity for savanna-like landscapes is one of the neurological quirks that we humans carried with us as we spread across the world. Another is the desire to collect curios, especially pieces of the past. We're a storytelling species, so perhaps these artifacts are anchors and touchstones for the tales from which we find our reality. Whatever the cause, this urge nearly destroyed all of the Florissant fossils. In the late nineteenth century, a rail line brought sightseers to the redwoods and shale. Within a few years, almost every petrified log and stump was gone and tourists had ransacked the shale along the train tracks. Several resorts occupied the site in the first half of the twentieth century, including one whose lodge was built on the hill behind the Big Stump. Lodge owners dug the mudstone from around the buried stump, giving tourists a better view. Builders broke and mortared petrified wood to make fireplaces and mantels for cabins. Business was brisk. Walt Disney came, liked what he saw, and sent a mechanical crane to carry a large stump to California. The well-traveled stump still stands in his theme park and befuddles visitors with a plaque engraved with geologic nonsense. Early scientific collectors knew more geology than Disney; they also dug more land. Using shovels and horse-drawn plows, the scientists hauled out large quantities of shale, most of which is now dispersed across museums in the eastern United States. In the 1970s the National Park Service acquired the Florissant site, and fossil collecting, except for small scientific digs, was forbidden. We now collect our curios by camera and sound recorder, or from the private quarry down the road.

The frenzy of collecting has left many marks at Florissant. In dozens of hours sitting at the Big Stump, hearing the chatter of visitors, I found that the most commented-on features of the site are not the redwood's gorgeous grain or the stunning sound of ponderosa pines but the two rusted saw blades that jut from the stump's upper half. Like knives left in a partly eaten cake, they lie in the vertical cuts that their teeth forced through the rock. The blades are snapped and stuck. Trapped, they have lain rusting ever since the 1890s, when they were abandoned after a failed attempt to divide the stump and ship it to the Chicago World's Fair. Despite the wooden scaffold and steam engine that supported and powered their work, the blades were no match for the size of this stone column. The rusted blades remain, unplanned monuments to the destructive effects of our desire to possess the past.

Saw blades may sometimes fail, but it is easy enough to pocket a stone, to remove the broken physical remains of the redwood. Finding and carrying with us the meaning of the tree is a much harder task.

An antique Kyoto kettle, heated over charcoal, hisses, evoking the sound of wind in evergreens. Our ears discern two groups of trees: one close at hand, the other far away. The kettle's walls click steadily as the vessel warms over the fire and, amid the sibilant trees, we hear a woman's footsteps. We're in Yasunari Kawabata's *Snow Country*, and the tension between intimacy and distance, heard in the landscape, gives voice to the book's central theme. The sounds also mark the start of the novel's final scenes: hearing trees mingled with human footsteps, Shimamura, the dissolute protagonist, turns away from human relationship and is engulfed by the cold night sky. He falls out of this world into a lonely void.

In Colorado's snow country, we also hear two evergreens. One near to us, a ponderosa pine living in our own time. Another, the redwood, sings from the distant past. In the ecological dissonance between these two trees there is also an opening to a void, a path to emptiness.

The petrified stump, a stony piece of flotsam carrying the memory of the past, reminds us of Earth's unnegotiable law. What exists today will not exist tomorrow. Climate change is one expression of this ephemerality. All the climate has ever done is change: cadences and glissandos of temperature and rainfall, sometimes bending slowly, sometimes screeching in jolts. This is the neverstill of rocks, air, life, water. Next to the petrified wood, the ponderosa cries in an igneous wind, prey to onslaughts of beetles or drought, caught in the change that humans have wrought. Downstream the effects of this change: slop buckets of glass shards and forest mud, thimblefuls of mountain slide.

No one marches in Washington to protest biogeochemical modulations or revolutions, yet I've joined thousands on the streets to protest the lack of adequate policies to slow human-induced climate change. What cause have we to divide relationships to the world in this way? If we choose not to inhabit a world of separation, if we believe ourselves to belong here just as much as pines and redwoods, where do we root our ethics? One philosophical consequence of the Darwinian revolution is to puzzle over this question. If we're made from the same substance as all other creatures, if our bodies emerged from the same naturalistic rules, why be concerned with another change in the climate, brought about by another natural process, the actions of humans? Our behavior is just as Earth-born as the geologic forces that ended the Eocene period or the actions of species that remade the atmosphere through their constructions and wastes. Life has continually reshaped cycles of limestone, oxygen, carbon, ozone, and sulfurous gases, sometimes with cataclysmic consequences for the planet's overall biological diversity.

Our secular and theistic traditions often answer ethical questions by positing a divide between *us* and *them*. God grants us special responsibility as stewards of our cocreated neighbors or, without God, we've acquired unique status through language, art, or technology. Such beliefs seem discordant with the unified ecological richness of the world. Dogmas of separation fragment the community of life; they wall humans in a lonely room. We must ask the question: can we find an ethic of full earthly belonging?

The answer depends, at least in part, on the kind of Earth to which we think we belong. If the world is a dance of atoms, regulated by physical laws and no more, then one answer to the question of an ethic of belonging must surely be ethical nihilism. In *Snow Country* Shimamura walks away from the people and the landscape to which he could have remained bound. He unearths himself, disappearing into the void, the remote physicality of the stars. Our two trees, living in such different climates, suggest a similar path. If we're a species made merely of atoms like all other species, no more and no less, it is a puzzle why we should believe that the human-caused climate change threatening ponderosa pines is an ethical calamity but regard the changing climate of the Eocene redwood forests as an ethically neutral phenomenon. This puzzle deepens if we turn the naturalistic lens on the origins of our own ethics. Many biologists claim that our thoughts and feelings of "ethics and meaning" derive only from the proclivities of our nervous systems. Our behavior and psychology developed by the process of evolution, as did the minds and emotions of all animals: no *us* and *them*, just different variations on evolutionary themes. If so, ethics are vapors arising from our synapses, not truths with objective validity outside our own minds.

Familial and group loyalty. Empathetic response to the suffering of others. Emotional attachment and assignment of "value" to endearing organisms. Biophiliac need to be close to trees. Concern for the preservation of our own kind. Human rights, animal rights, and intrinsic value of species. These are all deeply felt human beliefs, but outside our neural tangles do they possess any truth, any meaning? A different evolutionary path would have produced in us different genes, different ethical systems. Nihilism is perhaps therefore an attractive answer to the question of truly belonging to the physical and biological order: Our ethical beliefs are self-deluding dreams, "weak and idle themes." No matter if one ephemeral species burns the fossil remains of another, thus warming the planet a little. No matter, indeed, about anything at all, unless we care to divert ourselves with illusion.

Yet I seek something less fractured, an ethic that is fully biological yet does not walk us into Shimamura's starry, cold universe, empty except for self-constructed miasma.

A hint at such an ethic might be found with the girl in pink trousers who heard the "huge" sound in the ponderosa. She and her family were attending to Florissant with delight and unaffected ease. The girl heard the tree. The boy examined fallen ponderosa cones, peering between their open scales, then poking at immature cones on the tree. The parents noticed and pointed out the wavelike motions of wind on meadow grasses. The children read the Big Stump's signage without prompting and with a curiosity devoid of ostentation. They stood and admired the giant stone, remarking on its variegated colors. They remained at the stump far longer than the minute or two allotted in the walks of most visitors. This family was present. Like anyone starting a new friendship, they were listening and exploring. Theirs was the start, or perhaps the continuation, of a sturdy relationship with Florissant, a sensory, intellectual, and bodily opening to the place. The people formerly indigenous here— the Ute Indians and their ancestors—were forcibly removed in the nineteenth century, an act of violence that broke humanity's millennia-long relationships within this part of life's community. Much of the memory and knowledge embedded in these relationships died. The girl and her family were taking the first small steps in relearning part of what has been forgotten.

The family's attention to the particularities of Florissant seems at first to have little to do with understanding the ethical import of mud slides in the Eocene and in the present day. The family's behavior gives no direct answers to particular questions about the ethics of climate change. Instead, they may show how to move toward answers by engaging the community of life. From this engagement, or reengagement after cultural fracture and amnesia, comes a more mature ability to understand what is deeply beautiful in the world.

Ecological beauty is not titillating prettiness or sensory novelty. An understanding of life's processes often subverts these superficial impressions. A burn "scar" can in fact be a long-awaited renewal. The microbial community under our feet may be more richly beautiful than the obvious grandeur of a mountain sunset. In rot and scum we might find the slimy sublime. This is ecological aesthetics: the ability to perceive beauty through sustained, embodied relationship within a particular part of the community of life. The community includes humans in our various modes of being within the biological network, as watchers, hunters, loggers, farmers, eaters, story singers, and habitat for microbial killers and mutualists alike. Ecological aesthetics is not a retreat into an imagined wilderness where humans have no place but a step toward belonging in all its dimensions.

In this ecological aesthetic we might then root our ethic of belonging. If some form of objective moral truth about life's ecology exists and transcends our nervous chatter, it is located within the relationships that constitute the network of life. When we are awakened participants within the processes of the network, we can start to hear what is coherent, what is broken, what is beautiful, what is good. This understanding emerges from sustained incarnate relationship, becomes manifest in a mature sense of ecological aesthetics, and gives rise to ethical discernments that emerge from life's network. We transcend, at least in part, the individuality of our bodies and our species. This transcendence emerges from the earthly realities of the processes of life and remains agnostic on the question of whether gods or goddesses are involved.

Iris Murdoch, building on Plato, wrote that the experience of beauty was an "unselfing." In writing about this experience, she used the sight of a kestrel in flight as her first example. Although she claimed that both human life and the life of the kestrel had no purpose, being ultimately pointless, she argued that our experience of beauty in the kestrel was "patently a good thing" and a starting point for virtue and moral change. She did not expand these ideas in an ecological context, nor did she emphasize the depth that a sustained relationship with kestrels might add to our ability to find the birds' beauty. Not far from her home, J. A. Baker was conducting just such a practical and literary experiment, following peregrine falcons, becoming unselfed into the birds' lives. An ethic of belonging would merge and extend Murdoch and Baker, yielding sophisticated understanding of beauty's relationship to ethics fed by extended embodied experience. We unself into birds, trees, parasitic worms, and, sooner or later, soil; beyond species and individuals, we open to the community from which we are made.

The nihilist, and others, will argue that what is beautiful is an illusory high smoked out of the particularities of human sensory biases. As Hume wrote, "Beauty is no quality in things themselves: It exists merely in the mind which contemplates them; and each mind perceives a different beauty." But pause and consider the mathematicians. Theirs is an exact practice, one that seeks to find, if not objective truths, then at least the closest we can come to such. We trust this mathematics to keep airplanes in the sky, find new subatomic relationships, and hold the weight of the roofs over our heads. Before aeronautics, physics, and carpentry comes aesthetic mathematical judgment, developed through years of relationship with a particular branch of the discipline. Mathematicians use beauty as a guide. Elegance is one criterion of rightness or of steps on the path to rightness. Training and experience are necessary to see this elegance; only someone in deep relationship with a mathematical problem can discern such beauty. Paul Dirac, the founder of quantum physics and no friend of theism or mysticism, spoke of "getting beauty in one's equations" as a method of seeking fruitful insights. He claimed that mathematical beauty was in many cases a more reliable guide in physics than strict agreement with experimental results. Richard Feynman wrote that we make predictions about unknown areas of physics because "nature has a simplicity and therefore a great beauty" that is revealed through mathematics, the method that finds "the deepest beauty" of the world. Feynman also echoed Kepler and many earlier mathematicians in describing important equations as precious metals or jewels.

Mathematics, therefore, provides a precedent for using an aesthetic sense born of deep relationship as a guidepost to truths that transcend the human mind. We might do the same within the relationships that constitute biological networks. Someone who has listened to a prairie, a city, or a forest for decades can tell when the place loses its coherence, its rhythms. Through sustained attention, beauty and ugliness, in all their intermingled complexity, become heard. Unselfing through repeated lived experience is necessary because many biological truths reside only in relationships beyond the self. A casual visitor cannot hear these truths, still less someone visiting in mind only, applying abstract ethical schema from the seminar room. Even Hume might agree: "Strong sense, united to delicate sentiment, improved by practice, perfected by comparison, and cleared of all prejudice, can alone entitle critics to this valuable character; and the joint verdict of such, wherever they are to be found, is the true standard of taste and beauty." In ecology such a joint verdict must include the experiences of many species, perceived through the practice of unselfing.

This approach to ethics breaks the barrier between humans and the rest of life's community. If mature aesthetic judgment comes from extended relationship within an ecosystem, then other species are surely just as qualified for the work as humans. Further, ethical statements can be made about the changes wrought by any of the universe's members: humans, volcanoes, bluebirds, or rain. If there is objective ethical content in the tumult of biology and geology, it is present regardless of whether humans are standing by to judge. A mass extinction is a bad thing in and of itself, as will be the final obliteration of the Earth by an expanding Sun. But if ethics are human-made illusions dwelling in the subjectivity of nervous systems and not beyond, such claims are absurd.

No doubt what a raven, a bacterium, or a ponderosa pine sense in their worlds is radically different from what I perceive. These creatures also process what they sense in divergent ways. But such variations are not necessarily barriers to aesthetic and ethical judgment.

Beauty is a property of networked relationships that might be heard through ears of peculiar and multifarious design.

Ravens have internal, centralized processors, nervous systems and brains with some similarity to our own. They also live in social networks in which thought and intelligence reside, just as thought and intelligence reside within human culture. The ecological aesthetics of ravens therefore bear some relation to our own experience. We do not know whether ravens distinguish between mere prettiness and deep beauty or what this might mean for their sense of rightness or wrongness in the world. But nothing in biology bars ravens from making such connections. Parsimony suggests that similar nervous systems might produce similar outcomes.

The bacterium processes information not in any individual cell but through interactions within a soup of compatriots. The surface of each cell effervesces with chemical activity, the chatter of the group as signals pass from one cell to another. Bacterial intelligence is almost entirely externalized, contained within the chemical and genetic linkages of a community composed of thousands of cells of many species. These linkages are plucked, hammered, stroked, strengthened, or snapped by environmental change. Groups of bacteria signal, eavesdrop, and manipulate within this chemical conversation. Biologists call the networked decisions that emerge "quorum sensing." But this is no parliamentary meeting with nay/yea votes on single motions. Rather, the conversation is rich, is never ending, and results in nuanced changes in the chemistry and behavior of the community. Can the bacterial network be said to have an aesthetic sense that could make an ethical discernment? Not in any way that would be familiar to the experience of human brains. Bacterial aesthetics and ethics would be diffuse and strange. But perhaps no less true?

Ponderosa pine senses, integrates, weighs, and judges the world in a manner that combines external and internal intelligence. The ponderosa is networked to bacteria and fungi through every leaf and root. The tree also possesses its own hormonal, electrical, and chemical network. The trees' communicative processes are slower than animal nervous systems and they pervade branches and roots, rather than knotting themselves into brains. Like bacteria, they inhabit a reality alien to our own experience of the world. But trees are masters of integration, connecting and unselfing their cells into the soil, the sky, and thousands of other species. Because they are not mobile, to thrive they must know their particular locus on the Earth far better than any wandering animal. Trees are the Platos of biology. Through their Dialogues, they are the best-placed creatures of all to make aesthetic and ethical judgments about beauty and good in the world.

Unlike Plato, who sought through beauty to find invariant universals that exist beyond the caved-in mess of human politics and society, ecological aesthetics and ethics emerge from relationships within life's community. They are context dependent, but a contingent near-universality may emerge when many parts of the network converge on similar judgments.

The soughing wind in pines has carried messages across ages and cultures. The wail, whisper, dirge, and sigh of evergreens appear in great works of visual art, theater, poetry, and fiction in the East and the West. One of Ma Lin's famous paintings from the Chinese Song dynasty shows a scholar leaning in to a gnarled pine trunk; the painting's title tells us he is "quietly listening" to the pine. He peers into space in attentive puzzlement while a clear-eyed boy looks on. More than seven hundred years later, we're still leaning to the tree, trying to understand what we hear.

In *Snow Country* Shimamura's life pivots toward meaninglessness as he hears the sound of pines in the kettle. The pines, and the woman's footsteps that ring through the trees, drive him to turn away from this world. We too heard trees, near and far, and with them a girl's footsteps. Like the girl and her family, we can refuse to follow Shimamura in his flight, instead turning toward the sound, toward the trees. In the ethics of networked ecology, the principal practice and path is repeated listening.

The pink-trousered girl, if she continues to attend to the trees, will become the person to whom we should turn to make sense of the seeming dissonance of Eocene redwood and modern pine. Like Ma Lin's boy, standing to the side but perceiving more than his master, her open senses can guide us. Her words will emerge not just from her own being but from all she has gleaned from an unselfed experience within life's network.

Interlude: Maple

I. Sewanee, Tennessee
35°11'46.0" N, 85°55'05.5" W
II. Chicago, Illinois
41°52'46.6" N, 87°37'35.7" W

Maple I

I stand on the ridgeline of a house, my arms reaching up into branches of the maple tree whose trunk stands two meters from the front door. In one hand I hold a living maple twig. In the other, a square aluminum frame the size of my palm. Feet steady on the roof shingles, I ease a living branch between the metal struts, centering the thin-barked twig within the frame. A plunger descends from a capsule on the aluminum top bar, holding a small metal plate on the twig's surface. A weak spring inside the capsule pushes the plunger against the twig with a force as gentle as a breath. The pressure is so slight that the twig will grow unimpeded. Should the twig expand or contract, even by a just a fraction of a hair, the plunger's arm will transmit this motion to sensors in the capsule. The twig's skin now lies against the keen touch of a mechanical fingertip. A graph line on a computer screen reveals the metal fingertip's sensory experience, taking one measurement every fifteen minutes, year-round. Now it is late winter and the tree is leafless. No water flows through the twig from stem to leaf to air. The twig therefore lies quiescent against the plate. The graph's line is flat, disturbed only by the slight expansion and contraction of the frame's metal on sunny days and cool nights, and by the jostling of passing squirrel feet.

Maple II

"Hold these two blocks of maple wood and tell me which sounds better." He presses two slabs into my hands, each the weight of a heavy book. The blocks are hewn into wedges with surfaces rough enough to graze my hand. From each the luthier will craft a violin's back. For now, I hold seemingly soundless lumber, inert blocks. As I've been taught, I let my fingertips press into the wood, listening.

Maple I

The first week of April. Lime yellow flowers dust the maple tree's canopy. Peppercorn-size bells hang from filaments at the tip of almost every twig. A westerly breeze rocks the bells and pollen smokes from their lips. The twig that I'm watching, held against its sensor, ends with a spray of a dozen filaments, each with six pollen-shaking anthers under its bells. The branch from which the twig emerges has nearly three hundred twig tips. There are fifty such branches on the tree. A million or more anthers on the tree; insects know this well. Thousands of tawny wasps and black-green bees bounce on anthers, a *swir*, a hum, audible only when I climb into the treetop.

The sensor reports that the twig's diameter is mostly steady. But on sun-warmed mornings, the graph's line twitches down, then rises in the afternoon, a hint of water moving to flowers at the twig's tip.

Maple II

"It's this one, in my left hand." Then the analytic mind muddies my choice. Surely I felt no difference. Look at these wood lumps; they're the same. Skin, though, picked the better wood. As my hands moved, they sent tremors into the maple and my skin felt the answering reflections. The left-hand block had a shade more clarity on the hand.

Maple I

The second week of April. Last week's anthers are browned and falling, clogging the roof's gutters with tangled, woolly threads. Leaflets, each the size of a mouse ear, unfold from within cracked bud scales. Some leaflets spread from the twig's terminal bud, pushed out by the first motion of the elongating twig stem. Others splay from the paired buds on the twig's shaft.

The twig's sensor records an irregular flutter. At night the twig sometimes pushes outward by twenty micrometers, one tenth the thickness of this page. During the day the diameter judders down. The rhythm is jerky, unsteady, and, on sunless days, returns to the quiet of winter.

Daily, leaflets double in size.

Maple II

"Hold them up one at a time, tap, and listen. No, not like that. Left hand holds the top right corner, dangling the wood under your wrist. Tap the bottom left corner now. Tap with your finger pads. Your knuckles are deaf." Lodged in my fingers' epidermis, touch receptors waken.

When low-frequency vibrations run through my skin, they pulse against the tips of Meissner's corpuscles. Each corpuscle is a conical stack of skin cells enclosed within a thin sheath. Inside, a nerve twists through the layered cells. The corpuscles lie just under the skin's upper layer, located where the lightest of touches will reach them. When vibrations arrive, the nerve rouses and fires. The same vibrations stimulate disc-shaped Merkel cells that sit in my fingerprint ridges and in hair follicles on the backs of my fingers. When gentle pressure displaces my skin, even by a thousandth of a millimeter, Merkel cells sing. Higher vibratory frequencies provoke another set of receptors, the Pacinian corpuscles whose onion-shaped heads are embedded in the deeper layers of my finger skin. Each onion comprises dozens of concentric layers of membrane. A nerve sits at the center of the corpuscle, waiting for a trembling touch that arrives suddenly or penetrates deeply. Just below the skin's surface, spindle-shaped Ruffini corpuscles stretch horizontally within the skin and feel sliding motions or sustained pressure. Amid these sensory bulbs, disks, and spindles, free nerves snake throughout the skin, scavenging finger sound.

Like food or wine in the mouth, or words in a mind, touch has many waiting listeners, many dimensions. This community of receptor cells wraps its hearing into my nervous system, where it weaves into the fibers that stream from my inner ear. My mind struggles to tongue, to name, what the maple sounded against skin and eardrum. Two blocks: the same feel, the same tap. Or so my conscious mind said as it listened to hand and ear. Not a breath between them, yet something else spoke. The first block is bright, open, and lean muscled, quick. The second is very similar, yet tinged with granularity and turbidity.

Maple I

The third week of April. A summer tanager gleans caterpillars from the highest leaves of the tree. Between bouts of leaf probing, the bird mulls the air with its song. Immature maple fruits, miniatures of the samara helicopters that will fall later in the year, dangle fat and glossy from twig ends. The maple's leaves are as big as they will grow. Insect mouths have already shredded, rolled, and punctured many leaves.

Breezes, silent in early April, have found their maple voice, a flow of sluicing sand. Along the twig, another sound, one whose oscillations are ten million times slower than the susurration of leaves. Like an aorta pulsing with flow, the twig swells through the night as cells plump with water. After dawn the sun draws vapor from the leaf and the twig pinches inward, like a straw collapsing as thirsty lips pull on a drink. This in-drawing continues through the morning until, at noon, the twig is forty micrometers thinner than it was in the pre-dawn hours. On most days, by midday the roots have slurped, scavenged, and exhausted soil moisture. The upward flow of water ceases. To stem the hemorrhage of water into the afternoon air, leaves close their breathing pores and the twig's tightness gradually eases. As evening comes, water returns to roots and stem. The twigs' girth expands. Onto these rhythms the twig daily adds slivers of wood to its diameter, new cells, expanded cells. After a sunny week of vigorous growth, the graph's noontime trough is higher than the predawn peak of seven days ago.

Maple II

Scoops, finger planes, and chisels lie on a workbench. The luthier lifts two wooden pieces from a bed of shavings: a violin's back and its belly. The back has the sweet-edged nose nip of unfinished maple wood. The belly's scent is more acerbic and dusty, the smell of dry spruce. After the heft of blocks, both feel light as parchment. But parchment is a blotting mire for sound, the antithesis of the violin's clarity.

The violin's back and belly are delicate, precise parings of tree. They are like air to a flying bird. My thumb and forefinger startle at the speed and strength of sound's wingbeats. The luthier has given trees what Japanese carpenters call a second life, a life that can be as long and rich as the first.

Maple I

All summer long the forest throbs with the water-blood heartbeat of twigs. With sensors on other branches in the maple tree I hear how these heartbeats vary. Twigs on moribund branches—too shaded and low in the canopy—have weak pulses, daily cycles that barely stir under the metal fingertip. On sun-happy branches, systole and diastole surge and draw back, the forest's subsonic hum.

Maple II

"Here's the last violin my father made. It was not quite finished. I keep it here. Hold it." As we speak, the back and belly quicken, responding to every syllable. The curved skin of wood presses into air's caress and answers with a shiver.

Part 3

Cottonwood

Denver, Colorado
39°45'16.6" N, 105°00'28.8" W

A young cottonwood tree stands on the bank of a creek in Denver's urban center. The tree reaches only as high as my chest, a dozen thumb-thick stems sprouting from a clench of roots. The roots are lodged in a crack among a disordered pile of quarried stone blocks, part smothered with river sand. A municipal garbage receptacle stands on a concrete walkway next to the tree. On the other side of the tree, a meter-wide stretch of sandy gravel leads to fast-flowing, shallow waters where Cherry Creek runs into the wider, deeper South Platte River. The tree stands in the accumulated sediment in the concavity of the creek's arc. The cottonwood's companions in this narrow band of vegetation between walkway and water are shrubby willows, pointing downstream like the cottonwood, bowed by a long-sunk springtime flood. Shreds of plastic bag and willow stem are trapped in the cottonwood's lower branch axils.

On a warm afternoon, bicycle chains *twhir* as they pass the spattering sounds of wind-clapped cottonwood leaves. One or two bikes

pass every minute. Runners scratch a percussive beat on sandy concrete. Stroller wheels grind. Mayflies rise from the river's surface and
are plucked by chittering barn and cliff swallows. The birds sweep up
into the recesses of the Fifteenth Street bridge, where they have
daubed their mud nests onto metal girders. In the water dozens of
children shriek, whine, and whoop. A young man cannonballs into
the South Platte, a stinging *whump* of skin on water. He bobs, then
swims for the bank, his long black hair a gleam of shedding water
drops. Kids—black, Latino, white, Asian—float in the shallows, foot-
splashing from within inflated plastic tube creatures. On a rock under the cottonwood, a tattoo-vined couple watch, share a cola, and
laugh. Their toy dog, clipped into a yellow life vest, refuses to swim.
Cottonwood leaves yaw in the gusting breeze.

It is a sunny weekday in late summer and at least 150 people are
in Confluence Park, named for the union of the South Platte and
Cherry Creek. But the confluence here is of more than water.

Cherry Creek is the translated Arapaho name for the choke-
cherry, *bííno ni*, that grew on what are now paved walkways. French
trappers and traders gave the Platte its name, for the "flat" water of
its lower reaches in Nebraska. The tumble of vowels against conso-
nants in the Arapaho name, *niinéniiniicíihéhe*, is a better representa-
tion of the river's sonic and visual qualities in this more lively
stretch. Arapaho trails converged here and the riverbanks hosted
encampments of thousands. The exact size of these communities before contact with white violence and disease is unknown. The Arapaho were swiftly pushed out by colonists and, following the Sand
Creek massacre in 1864 and relocation to Oklahoma, living Indian
presence here was obliterated. Today the Arapaho are acknowledged
in street names, historical signs, and wall art, but not in restitution
of land and power. Encampments continue, though. Dozens of
homeless people sleep on cardboard in the willow thickets. An outdoor recreation store overlooks the park and a sign inside reads "we

love to get outside and play, and we know first-hand the importance of quality outdoor gear."

Denver's nineteenth-century colonists arrived by river and on trails alongside the banks of the South Platte. They built, as had the Arapaho, at the confluence. Many built *in* the confluence, erecting houses and business offices on the sandy open ground alongside the river. Flash floods repeatedly smashed through these monuments to riverine ignorance. When thunderstorms hit the upstream reaches or snowmelt came suddenly, the plash of the creek turned to a roar powerful enough to carry downstream, in various decades, City Hall, Denver's first bridges, and the three-thousand-pound printing press of the *Rocky Mountain News*. Dozens died in these floods. For decades eastern immigrants rebuilt in the pathway of the water until, in the early and mid-twentieth century, upstream dams dampened the floods' vigor and building codes edged structures away from the riverbed. Today Confluence Park is set lower than the surrounding apartment buildings and stores, a *Y*-shaped catchment for the rare two-meter surges that dams cannot contain. On most days, when the water runs low, the park's design has an unintended acoustic consequence. Although Interstate 25 cuts through within ten minutes' walk and some of the city's most busy streets demarcate the park to the north and south, the clamor of tire and piston largely overshoots the basin, opening a space at the center of the city for the purl of water and the voices of children, cottonwoods, and birds. Here the city is a low drone spiked with sirens and motorbike pipes. The river cuts a middle way between these extremes of rhythm, loudness, and tone: the weir is a bass roll with splashy grace notes, its steadiness unifying the tapered riffs of animal and plant voices.

Cottonwood depends on temperamental flow. Floods scour the high ground along the river, leaving moist, sandy beds for the trees' seeds. The "cotton" that wraps these seeds carries them on wind and water.

Only on bare ground can the seeds germinate and grow. In established vegetation the flecklike seeds are too puny to compete. When the river drops, the seedlings squirm roots into bare sand, chasing the falling water table. The tree grows upward too, but the pursuit of water is the seedlings' focus. After a few weeks, the shoot may be as tall as a finger, but the roots are arm-length deep. If they fail to keep up with the ever-sinking layer of sandy moisture, the young trees wither on the desiccated riverbank. Seedlings that germinate when the river is low are usually washed downstream on the first flood. Only those that start life high on the riverbank, on the ebb of a flood, grow into mature trees.

Suppression of floods was the primary objective of Denver's first dams. Dams change the pulse of the river from an irregular beat, with great floods coming at unpredictable intervals, sometimes years apart, to a regulated, even flow, punctuated by regular washes from dam discharges. Below dams cottonwood often disappears from riverine forests, replaced with Eurasian tamarisk, a species that thrives in the new water rhythms. Around Confluence Park there are few young cottonwoods. The river is edged with tended lawns and concrete walkways. High water here now departs with few seedlings in its wake. In the slivers of unmanaged vegetation and riverside riprap, though, the old order persists. Seedlings find the spot moist enough to grow yet high enough to stay rooted. Park managers lend a hand, planting young trees at the edges of mowed lounging areas, replacing the force of the river with the foresight of humans.

At the park I push the curfew to its edge. No overnight camping or loitering is allowed, although the homeless find ways around the rules. Adonis tinkles as he packs his wineglass, bottle, and book into a leather shoulder satchel, then mounts his titanium road bike. Water wrestlers, three teenage boys in Mexican flag T-shirts, shake themselves dry and jostle on the bank before slapping their flip-flops up the concrete ramp to the bridge. A mother pleads one more photograph from the fussing, petticoated infant posed on a rock pedestal in Cherry Creek's bubbling swirl. An old man grunts as he rolls down the legs of his shorts. He shoulders a white shirt as he stands, his sun splay on the bench over for the day. A cat carrier's door clicks shut as a lawn-wandering boa is caged. A muscular, goateed man gently lifts the boxed snake and walks to a bus stop. The bridge's bilious security light fizzes with an unsteady, insectile spark. Their sandbar freed, mallards preen and chuckle under cottonwood stems. Then, a bark I know from coastal marshes, *rwonk*: a black-crowned night-heron glides above Cherry Creek and drops into the jumble on the rock island in the middle of the South Platte. The heron edges out of the rocks and treads its long toes to the water's edge. There it stares down into the reflected streetlight shine of its silver plumage. I crouch behind the cottonwood, the better to observe without alarming the bird.

Over the next two years, I'll learn that I had no need of stealth. The night-heron is indifferent to human fluster and palaver. The whir of speeding commuter bikes along Cherry Creek or babble of children near South Platte's rocks do not draw the bird's red disk eyes from their fish-dagger stare. Here are the Galápagos in downtown Denver: birds in whom the inner voice of fear has quieted. Annie Dillard called the approachability of Galápagos animals "pristine ignorance," their welcome when they investigated her "the greeting the first creatures must have given Adam." The islands were a world where animals had not yet been defiled by the touch of fallen humanity. The Denver night-heron confounds this allegory. These same Edenic qualities live in the heart of a city that humans not just touched, but built.

I return to the tree in winter and find a haze, a dome of urban incense. On cold, sunny days millions of spinning tires are censers, flicking road salt particles to the air. There they join exhaust fumes and ozone to make a cloud of pollution over the city. From a distance, Denver genuflects its smoky offering at the foot of the glass-aired Rocky Mountains. The bright, diverse paintwork of cars is unified by a veneer of gray brown powder. Tree trunks are smutted with dirt-rusty taupe, the color of moles and mined earth.

Denver's road salt comes from Utah, an old seabed hauled from its sanctuary through kilometers-long tunnels the height of houses. During an average winter, road crews scatter nine thousand kilograms, ten U.S. tons, of pulverized Utah rock salt per mile of Denver road lane. Downtown, to reduce dust, the crews use a spritzed brine of magnesium chloride. Two decades ago the cloud was thicker. Then, salt and sand were sprinkled at three times today's volumes. Breathing was a geologic experience, wet alveoli of lungs muddied by hovering rock strata. Now road managers use salts in a much more efficient manner, but even today power lines are sometimes shorted by salt accumulations. The "perpetually brilliant" sunshine and "tonic, healthy" atmosphere promised by Colorado's nineteenth-century boosters are dulled by our collective desire for rubber to meet asphalt with speed and surety, whatever the weather.

Snowmelt and rain clear both streets and air. But terrestrial clarity comes at the price of aquatic turbidity. The South Platte and Cherry Creek receive much of the salt, sand, and silt. Their waters, inflamed by runoff, are the phlegmy cough of the city. The flow in front of the cottonwood, usually clear as tap water, runs opaque and stained when winter snows melt. Denver's waters briefly become mine tailings.

A range map of the plains cottonwood species is an irregular oval in the center of the North American continent, hundreds of kilometers from any coast. Yet droughty soils have adapted the tree to periodic inundation by salt. Cycles of shallow rain and desiccation draw salt from deeper layers of soil. Every shower dissolves the soil's salt, then the sun moves these solutes higher as water is pulled by evaporation and the capillary grip of soil particles. A thorough soaking will leach salt away, but abundant rains are infrequent over much of the cottonwood's range. The ancestors of western cottonwoods therefore have some experience of salty soil, and the survivors passed this knowledge to the present generation. The trees cannot match sabal palm for their tolerance, but cottonwood cells can sequester salts within compartments, produce defensive chemicals to buffer themselves against the water-drawing power of salt, and grow roots that burrow lower than salty superficial layers of soil. Cottonwood roots also fuse themselves into salt-tolerant fungal networks, tapping the water, nutrients, and defensive chemicals of their partners. Like the ponderosa pine, the aboveground cottonwood shoot is the smallest part of the tree, a flagpole raised by an underground community.

Animals in the rivers and creeks have likewise inherited some resilience from their ancestors. But there is a limit. If chloride, magnesium, or sodium from road salt becomes too concentrated, fish and aquatic insects are sickened or killed. Billows of sand and silt can sink or smother wads of snagged dead leaves and mats of algae, burying the food that sustains the aquatic community. Trout get much of the attention in these waters, but their lives depend on algae-grazing and leaf-munching insects. These downstream effects of road treatments are part of what caused Denver's road managers to change what they loaded into salt trucks. The old sand/salt mixture sluiced far more particles and salts into waterways than do the newer Utah mine salts and brined magnesium chloride. Denver has a goal of returning every waterway in the city to a state in which fish can thrive. Thoughtful road management is one reason that people have dusted off their fishing poles along some Denver waterways. Other creeks await further progress. Now bullhead, catfish, shiner, chub, sunfish, dace, suckers, and even some trout live at the confluence of Cherry Creek and the South Platte. Unlike decades ago, line-swinging anglers are often seen standing in the water's flow.

With healthier rivers and creeks come new dangers for urban trees. Early in the first winter of my time with the cottonwood tree, I walked to the South Platte and saw the municipal trash can standing alone. All the tree's trunks were gone. I rummaged in the willow thicket and found ankle-high stumps. On each, pencil-thick grooves etched the slope of an angled cut. A spill of cottonwood chips circled the amputee. A few willow stems were nipped. Beavers had felled the tree's trunks and dragged them to their lodge downstream in the South Platte. City workers had finished the rodents' work, tidying with smooth lopper slices the skinny trunks that the beavers had ignored.

By the next summer, the tree was taller than the year before, just over two meters of multistemmed growth. In October the beavers returned for their winter provisions. They leveled the tree once again. Next spring: new cottonwood sprouts. The chisel-toothed rodents are rough managers of their trees. Most human foresters would not approve of such hard-driving, short-term rotations for a coppiced woodland. The cottonwood, though, seems to keep ahead of the game, growing a little taller each year. If the beavers let up, the tree would get big enough to undermine and crack the pavement. Such unruliness would likely cause park managers to remove the tree. The beavers' vigor, for this one individual cottonwood, therefore might ensure a longer life.

My conversations with the people who clear the snow from the paved river pathways, keep the trash cans empty, and tidy the discards of human visitors confirm that beavers live in many of Denver's urban creeks and rivers. Ted Roy, who has worked for the city for more than two decades, reveled in listing for me the animals that he sees on his rounds: beavers, coyotes, muskrat, foxes, hawks, snakes, bears, and "penguinlike birds," presumably night-herons. What gave him particular pleasure was seeing how much change had happened during his time with the city. Denver's waterways now host more wildlife, have better facilities, and are visited by many more people. Mr. Roy, riding in a municipal truck piled with garbage bags, is part of the river's memory and intelligence. His guffaw and backchat in the cab are the sounds of water wisdom, Baker's *Peregrine* translated to city life.

To better understand the cottonwood in Denver, I followed the South Platte just over one hundred kilometers upstream, to Eleven Mile Canyon in the mountains. On a late-summer afternoon, a young American dipper stands on a granite boulder in the river's headwaters and shrills a repeated, jabbing note. The bird's parent toadclimbs from the water's tumult, feathers shedding mercurial liquid, and twists a clump of mayfly nymphs into the squalling youngster's beak. The begging recommences before the adult has time to turn and submerge, grip footed, to its work on the river bottom. The dipper needs its crampon feet and finlike wings. Here the South Platte guns through its raceway of billion-year-old granite, a teenage river on a smash, ram, and whomp of a run from its elderly parents. The ruckus jams all sound from ponderosa or willow. Only the newly fledged dipper outbawls the river, the bird's call a high note vaulting the water's roar.

Late summer's fecundity is all around. On the meadows that tumble down slopes to the water's edge, mule deer does browse in the company of their brawny offspring, now outgrown the spots of fawnhood. A merganser duck sits with her brood next to a water riffle below the bouldery rapids. Seed-fat grass heads line the trails; canyon walls droop with cones. In air spiced with pine and river spray, the only sounds are of birds, water, and wind. Ah, the mountains. Here, John Muir tells us, after we have "bathed in the bright river, sauntered over the meadows, conversed with the domes, and played with the pines," we may finally shake "the last of the town fog" from our bodies and minds.

The only other humans are fly-fishing in the calmer reaches of the river. Some stand in national forest waters, others behind reflective metal advertisements, Private Fishing, No Trespassing, No Parking. The fishers cast their lines with arms encased in the UV-resistant, breathable fabric of well-engineered shirts. Many-pocketed vests hold the fly box, nipper, zinger, hemostat, nail-knot tool, fly-floatant powder, tapered leaders, and tippet spools. Their hats are wide brimmed, sturdy but foldable, and water sandals or wading boots keep feet as steady as waterbirds' on the uneven river bottom.

I'd guess that each trout fisher has about one thousand dollars' worth of kit. My clothes are no match, but the sound-fishing and light-luring electronic equipment in my backpack cost as much as their gear. We have the leisure to take a day away from jobs or families, money for entrance fees and gasoline, and cars reliable enough to climb from the plains to mountain canyons. All of us are male and seem to have a few decades of employment savings in the bank. And, using Ta-Nehisi Coates's pithy summation of race in early-twenty-first-century America, we all believe ourselves to be white. Formerly our seeming unity would have been sliced into hierarchies, "Catholic, Corsican, Welsh, Mennonite, Jewish." But now we white dukes, inheriting privilege through birth, go to the woods and streams and there, like Shakespeare's duke, "exempt from public haunt,/Find[] tongues in trees, books in the running brooks,/Sermons in stones, and good in every thing."

The same trees and stones have other tongues, other books and sermons.

Writing of her family road trip through the open spaces of the American West, Judy Belk recalled her son's reaction on first hearing of the trip. "Four black folks from Oakland" on the Montana back roads seemed, to him, "nuts." His was one expression of what Carolyn Finney calls "geographies of fear." American history, combined with the current state of racial inequity, reserves the feeling of wholesome ease in the outdoors for only a small segment of humanity. As an older white man, I approach the woods, the river, and uniformed, potentially armed rangers in a very different context from that of a black teenager. "Don't bird in a hoodie. Ever" is one of J. Drew Lanham's "9 Rules for the Black Birdwatcher."

The woods, creeks, and mountains are where many have disappeared. This too is the forest unseen, unheard. The lonely creek is where white men dump the bodies of those they have killed. The trees are hung with Billie Holiday's "strange fruit." The "outdoors"—fields, forests, and green spaces—carry with them memories, and present-day threats, of violence. When the National Park Service's Bill Gwaltney told his family that he planned to be a ranger, his father, who had lost friends to the noose, replied with a warning: "There are a lot of trees in those woods, and rope is cheap." The journalist and mountaineer James Edward Mills calls the legacies of past and present dangers a "cultural barrier forged in social memory." As a result, he reports that he is often the only black person in attendance at conferences and meetings about outdoor recreation.

It is not just racial injustice and violence that produce geographies of fear. A recent survey of scientists found that outdoor research sites were "hostile field environments" where 26 percent of women, compared with 6 percent of men, had been sexually assaulted. Red Riding Hood is partly a map of the geography of violence and fear. The tale also reinforces patriarchal cultural norms: girl, to be safe, don't wander into the forest or a man will have to rescue you from other men. Cheryl Strayed could walk the Pacific Crest Trail, in part, by telling "myself a different story from the one women are told. . . . I willed myself to beget power." Terry Tempest Williams, reflecting on her experience of human evil in the mountains, describes the process of "growing beyond my own conditioning." It is not, she writes, the lips of princes that will oppose and reshape geographies of fear, the "things that happen to young women in the woods," but "our own lips speaking."

The water of the South Platte flows through Eleven Mile Canyon in one channel. But several rivers are present here.

In the national forests and national parks of the United States, cultural geologies, the processes that create geographies of attraction and of fear, have been exclusionary from the start. These institutions were born from philosophies of Nature that reveled in the imagined supremacy of whiteness and masculinity. Muir, the leading advocate for the national parks, praised the "brave and manly and clean" mountaineers, men superior to the people in "crowded towns mildewed and dwarfed in disease and crime." A strong-willed white man, Muir believed, "would easily pick as much cotton as half a dozen Sambos and Sallies." Muir's Indians were "dark-eyed, dark-haired, half-happy savages" leading lives "strangely dirty and irregular . . . in this clean wilderness." Gifford Pinchot, founder of the national forests, was an avid supporter of the eugenics movement. He compared "races" of people to separate species of tree, "pines and hemlocks, oaks and maples," each human "race" living "in certain definite types of locality . . . in accordance with definite racial habits which are . . . general and unfailing."

Aldo Leopold used slavery in Homeric Greece as an example of an ethic that humanity had outgrown, but he treated the racial injustices of his own day with silence or ambivalence. Writing in 1925 at the height of Jim Crow, he argued that wilderness must be "segregated and preserved." Even as government policies were forcibly assimilating the American Indians of his time into white culture, he wrote that the "supply of wilderness" was unlimited when the Pilgrims landed.

These attitudes are enshrined at the entrance to the American Museum of Natural History in New York City. To enter we pass a twice-life-size equestrian bronze of Theodore Roosevelt. The statue's message is of unambiguous white superiority. Two men stand just behind the rider, both half naked, unlike the well-clothed president above them. The heads of the black man and the American Indian reach only as high as Roosevelt's buttocks.

It is perhaps, then, not surprising that the *Negro Motorist Green Book*, a publication designed to help black travelers avoid "difficulties" while vacationing in a segregated United States, makes little mention of parks and forests but instead lists private homes, hotels, and restaurants in cities. "Green" was the publisher's name, not the color of the recommended destination. In the 1949 edition, the closest safe hotel was sixty miles away from the environmentalists' darling, Yosemite. This despite the fact that black "Buffalo Soldier" cavalrymen were the guardians and caretakers of the Yosemite Valley and other western scenic areas before whites took over as these lands became national parks.

Century-old photographs of the Colorado Midland Railway, the mountain-climbing track that brought sightseers through Eleven Mile Canyon from Colorado Springs, show only white faces among the passengers, with an occasional black railroad worker. The South Platte may be a young river, but its channel is an old one, running through cultural granite.

Back in Denver later in the year, I walk on the South Platte shore on a rimy December morning. My boot soles are like twisting pepper grinders as they break the frost welds among sand grains. At the water's edge, an ice shelf juts over the lapping river wavelets, concentric milky circles marking its overnight growth. I step too close and part of the formation shatters, a window break that alarms the mallards, gadwalls, and hooded merganser ducks cruising the waters of the confluence. Then comes another startle of birds, a slap-panicked sound that resolves to a whistled hiss as one hundred pigeons bolt from their roost on the bridge. An adult bald eagle rows its dark wings with easy power. It has no interest in the corkscrewing silliness of pigeons but swings its head to eye the flat waters downstream of the weir. The eagle sees no stunned fish so continues its path, following every bend in the water. I hear the *whoff* of its wings as it punches a little extra power to clear the high struts of the bridge.

Gulls and Canada geese follow the same aerial path along the South Platte. Gulls snatch glances at the same fish-pregnant water as the eagle. Geese keep their eyes on more distant targets. Water from the irrigation sprinkler is the goose's Moses, come down from the mountains through reservoirs and pipes, opening a promised land. Half of Denver's water is used for watering ornamental plantings. In the sun-leathered parch of the western plains, Denver's lawns and the suburbs' well-landscaped office complexes are all that grass-grazing geese could ever desire: impounded water, thousands of hectares of fertilized and watered grass, and shrubbery in which to secrete their nests. The sky is seldom missing a skein, especially in winter, when flocks of residents and winter visitors use rivers and creeks as guides among the many feeding opportunities.

Humans too once again follow the river. By building more than 130 kilometers of walking and biking trails within the city, most of which follow waterways, Denver has aligned the movements of people with many of the other animals that inhabit the city. This confluence creates more than convenient or pleasant places for people to commute, play, or relax. People who are present along the river tend to become advocates for the river.

When human movement patterns start to realign with the patterns of other species—eagles, mayflies, geese, muskrat—our awareness rejoins the community of life into which we are born but which our built environments too often hide from us. In this unity of flow and bodily movement, belonging is no longer abstraction but is manifest through living choreography. The choreographer, though, is not an individual but the relationships among a multitude. The river is not a passageway for lifeless water molecules but is a life-form. I hear the Amazonian Sarayaku activist's words: *Rivers are alive and sing. This is our politics.*

Humans are part of this multitude. The South Platte and Cherry Creek flow from many upstream impoundments and diversions. The spreadsheets and management plans of Denver Water affect every drop of the river's flow. Do these manipulations by humans tame the river, somehow draining it of wild nature? No. The hand that writes water-management plans, the page or screen on which words appear, the engineers who devised dams, and the flow of the South Platte in the city are as wild, natural, and at home in this world as the waters within the upstream, federally cordoned "protected" area. We too are nature. Unsunderable.

To believe otherwise is to impose a duality on the world. The South Platte runs through a land created by this fissured imagination. The river gathers its first waters from mountain national parks, forests, and wilderness areas. For some people these areas are places for a grand escape, sacred groves in which to visit Nature, and the last refuge of imperiled ecosystems. For the indigenous and other peoples who were removed and barred from reentry before the federal government enacted "protection," the same areas are postapocalyptic landscapes. Cormac McCarthy's *The Road* runs through each one: Trails of Tears leading out of dehumanized lands. The Wilderness Act of 1964 preserves lands in their "natural," "primeval" condition where "the earth and its community of life are untrammeled by man." Indigenous communities in other parts of the world see the consequences of this philosophy that excludes "man" from the "natural" community of life. The Sarayaku oppose national parks in Ecuador, knowing the endgame of that idea. They prefer the term "living forests," where Life is understood to include people and the knowledge that dwells within people's many relationships with other species.

In unpeopled mountains the South Platte has its headwaters. Then the river flows to the city, where it encounters another manifestation of our philosophy of nature: pipes dumping effluent. When we believe in duality, we create duality in the world. If we think that the city is unnatural, then it follows that urban river water has fallen from its natural condition. Being already "trammeled," the water may then serve as a garbage chute. The corollary of the depopulated, protected "natural" area is, then, the industrial dump. By the 1960s, downstream of the mountain parks, the South Platte in Denver was bermed with industrial waste, scrapped cars, and heaps of castoffs from a rapidly growing city. Factories piped untreated waste directly into the waterway.

Once established, a binary landscape of nature and nonnature reinforces itself. As the contrast between wilderness and reckless development grows more striking and alarming, the need for "wilderness" appears to grow while the rest of the landscape gets seemingly ever more unnatural. In such a world, cities are disdained by environmentalists but unpeopled parks, forest reserves, and designated wilderness areas are lauded. As the landscape's duality grows, it gets harder to perceive that humans belong in the world.

Hostility to cities runs deep within environmental, agrarian, and scientific traditions. Thomas Jefferson wrote that "mobs of great cities add just so much to the support of pure government, as sores do to the strength of the human body." Virtue resided in white, rural "husbandmen." Muir encountered Nature when he escaped "intercourse with stupid town stairs, and dead pavements." Aldo Leopold's "land" includes "soils, waters, plants, and animals" but no collection of human abodes. Indeed, for Leopold "man-made changes are of a different order than evolutionary changes" resulting in diseaselike disorganization. Within academia the ecology of cities was, until the last two decades, not a topic of much interest to ecologists, even though the name of the discipline, *ökologie*, a German word coined by the nineteenth-century biologist Ernst Haeckel from the Greek *oikos-logia*, means the study of our dwelling place. Only in 1997 did the U.S. National Science Foundation add any urban areas to its flagship Long-Term Ecological Research program. Even today most biological field research stations are in areas remote from cities and towns.

The belief that nature is an Other, a separate realm defiled by the unnatural mark of humans, is a denial of our own wild being. Emerging as they do from the evolved mental capacities of primates manipulating their environment, the concrete sidewalk, the spew of liquids from the paint factory, and the city documents that plan Denver's growth are as natural as the patter of cottonwood leaves, the call of the young dipper to its kin, or the cliff swallow's nest.

Whether all these natural phenomena are wise, beautiful, just, or good are different questions. Such puzzles are best resolved by beings who understand themselves to be nature. Muir said that he walked "with nature," a companion. Many contemporary environmental groups use language that echoes Muir, placing nature outside us. "What's the return on nature?" asks the Nature Conservancy. "Just like any good investment, nature yields dividends." The masthead of the Royal Society for the Protection of Birds, Europe's largest environmental group, promises that the organization is "giving nature a home." Educators warn that if we spend too long on the wrong side of the divide, we'll develop a pathology, the disorder of nature deficit. In the post-Darwin world of networked kinship, though, we can extend Muir's thought and understand that we walk *within*. Nature yields no dividends; it contains the entire economy of every species. Nature needs no home; it is home. We can have no deficit of nature; we are nature, even when we are unaware of this nature. With the understanding that humans belong in this world, discernment of the beautiful and the good can emerge from human minds networked within the community of life, not human minds peering in from outside.

It is midday in August and although the park's trees offer shade, most people sit or flop in the open. I lack their western toughness and park my plummy skin under the cottonwood, now two years into its annual beaver regimen of pruning and resprouting. Fourteen stems grow from the root fist, five of them over two meters tall, enough for a person-size pool of shade.

From my riverbank seat, I gaze up through the lime of cottonwood leaves. Each one hangs from a straplike petiole. The planes of leaves and straps are perpendicular, so when the leaf moves, it wags from side to side. Unlike the broad leaves of other tree species that bounce like hands petting a dog's head, cottonwood leaves sway like hands wiping a window. Its cousin, aspen, does the same but with a more furious rubbing motion. Puffs of wind set the cottonwood tree tapping, hard leaf edges batting one another. Stiffer wind brings slaps as the waxy leaves strike with glancing blows.

These are the sounds of a fast-growing cottonwood tree. Despite the heat and low air humidity, the slap of the leaves reveals their full hydration. I'm thirsty and feel an inner shrivel, but the cottonwood leaves are fat with water, their discourse with the air full of wet vigor. Roots, now likely ten or more meters long, plug into many layers and patches of soil, diversifying the supply chain to ensure a continuous and abundant flow of water. This cottonwood grows in a manner that is as close to hydroponic farming as one can find outside of a greenhouse. Unclouded sunshine illuminates the tree on most days. Water percolates around roots, keeping them moist at all times. A slow drip of dissolved nutrients comes from the river and the leaching of higher layers of soil, including runoff from a fertilized lawn. In such abundance a tree should spread the light throughout its body, maximizing the flow of energy to its cells. Cottonwood leaf flutter achieves this end. Motion in the upper branches gives the topmost leaves a break from the sun's overwhelming power while opening a flickering supply of photons to lower leaves. The whole tree feeds.

The Confluence Park cottonwood's annual lancing recovery from beavers demonstrates its vitality. Unsurprisingly, cottonwood and its relatives are the favorites of geneticists breeding fast-growing trees for biofuel plantations. Along the river the cottonwood also is the favorite of leaf-chewing insects, nesting birds, and shade-seeking mammals. Without it riverbanks lose a keystone that holds much of the rest of the community in place. Unless upstream dams are managed to mimic the flows that nurture young cottonwoods, many species decline or disappear. Fortunately, this management strategy is now, very slowly, replacing the old ways, which focused on human needs only.

As the afternoon Sun blasts its heat onto the park, the metal receptacle behind me starts to ripen and release its trash-can smell. Although the aromas of foods, soils, and forests vary across the globe, public trash receptacles converge on one aroma, humanity's sensory common denominator. Robust undertone of apple rot. Fecal grace notes, no doubt from dog baggies here in fastidious Colorado. The prickly odor of microbial mats on the metal bottom, urban stromatolites. The cottonwood inhales this mix through its leaf pores and white slits on the soft skin of its green trunk. What the tree makes of this no one knows, but surely some of the odoriferous molecules bind to its cells, waking strange plant thoughts. What I make of the smell is clearer: time to move and, in this heat, take a dip.

The first lesson of my swim is that watery confluence takes time. Cherry Creek is pleasantly tepid. The South Platte punches the breath out of me when I reach its flow. Only after several minutes' downstream swim do the waters merge. My skin is learning Colorado hydrology. I'd studied the maps, but full immersion brings home the lesson. The South Platte comes from the mountains, shock cold. Dams held and warmed the water, but cold water sinks, so any reservoir release from below the superficial water layers retains a chill. Cold water holds abundant oxygen; insects and fish thrive in the gill-pleasing river. Cherry Creek has its headwaters on the plains at Castlewood Canyon, a slice of wetness in an otherwise dry landscape. The creek then runs in a shallow channel through much of Denver and its outskirts. The creek's origin and pathway are over hot rocks and concrete. At Confluence Park little kids always choose to paddle in Cherry Creek, their feet bathed in warm water.

An exfoliation of my knees and elbows brings the second lesson. The South Platte's bed is a tumble of scoured rock. This creates agony for limbs kicking and pulling against the water's sweep, finding stone where water should be. Teenagers bold enough to run the upstream rapids without a tube or kayak have great fun, until they start their rock-strewn swim back to shore after the ride. Their curses are manifestations of a river that retains its power. In impounded rivers, silt sags out of the drowsy water, smothering water-carved rock with soft mud. Not so in the South Platte. No wonder mayflies rise from the surface at Confluence Park. The favored rocky habitat of mayfly young is in abundant supply, at least in this stretch of the river.

Cherry Creek is bedded with sand, some of it from the slow erosion of tributaries in the fields and canyons but much of it from the bleed of soil and subsoil that accompanies the building of any city. As I walk back to the tree from my swim, I step on deposits that likely originated in jackhammered city street repairs, land cleared for a mall, and the combined trickles of dozens of housing subdivisions.

Yesterday I would not have stepped into Cherry Creek. A thunderstorm to the east turned it to a turbid, choppy speedway. Today the creek's bed is marked with the scallops and curves left by the surge, meter-long aquatic dunes running crosswise to the current.

With the silt and sand came swarms of living cells. Whatever went down the city's storm drains flowed through the creek, including *Escherichia coli* bacteria, the denizens of warm-blooded intestines. Denver's Department of Environmental Health monitors concentrations of these fecal bacteria and posts the results to a map. Although *Escherichia coli* itself is only sometimes dangerous, the species is easy to monitor and serves as an indicator of the many other pathogens that rain carries from streets and escaped sewage. A click on my social media account tells me whether it would be wise to dunk in this effluvium. After a city-washing rain, Cherry Creek is tagged on my computer screen with a red pin, meaning that bacterial counts are above the safe limit for recreation. When runoff from the storm abates, the pin disappears. The South Platte sometimes also receives high scores, depending on the condition of urban tributaries and storm-water drains. That either waterway is ever swimmable, let alone swimmable for a large portion of the year's bathing suit days, is a revolutionary change from the past.

As the river becomes more friendly to human and nonhuman life, new sources of *Escherichia coli* arrive. One hundred Canada geese loafing upstream of Confluence Park will, in one day, extrude from their active cloacae more than ten kilograms of goose droppings. A bulky animal that maintains itself by tearing at grass has a gut with a profligate rear end. On some days one hundred is a low goose count. The wildlife on Mr. Roy's list all make their own contributions to the flow.

Unhoused people join the geese. The South Platte's dense willow thickets make good bedding places for those whose lives have become untethered from sewage pipes. Some are homeless. Others define home not as a roof but as the community that lives in interstitial spaces of the city. For many of these members of Denver's human population, the river and its parks are attractive places to include in their routines. Without access to toilets connected to the municipal sewage plant, the bushes along the river must suffice. No matter how careful people might be, water quality suffers after rains.

As it was for the Arapaho and the first settlers from the East, the river is a gathering place and a camp. At sundown young travelers gather on the knoll just north of the cottonwood tree, an ingathering to greet, share, and plan for the night. In the morning, under the Fifteenth Street bridge, a gray-bearded man who had been "on the road for many years" told me that he started each day at the river with a rubdown for his body and a wash for his clothes. Like most of the travelers I spoke to at the river, he was happy to share some of his story but held his tongue about routes and sleeping places. Self-preservation requires circumspection. A young couple in smart hipster clothes, seemingly indistinguishable from other seventeen-year-olds, broke camp from under a cottonwood higher on the riverbank. The river was great, they said, much safer than many other parts of town. But settle in any one place for too long, and the human predators will find you. The geography of fear followed the river from the mountains and takes a new form in the city.

In winter the number of people who are sleeping outdoors becomes evident. Cottonwoods drop their foliage, revealing beaten-down leaves and cardboard-mattressed nesting sites all along the South Platte. Officially, a camping ban is in place, and police will move people on. But Denver's policies have vacillated. Portable toilets next to the river kept the water in better condition but attracted more sleepers to the park. Then the toilets were removed with the hope that the park's magnetism would diminish. A 2012 citywide ban on sleeping outside with "any form of cover or protection from the elements other than clothing" criminalized homelessness but has been unevenly enforced. A survey of Denver's homeless found that three quarters had been turned away from shelters for want of space, despite increases in the number of beds. One third of those surveyed in winter said they had tried to circumvent the city's ban by sleeping in their clothes only, with no cover. Such exposure must be hard in Denver's winter, but it meets the letter of the law. Denver's beavers and other river rodents now have better lodging than the river's people, an urban novelty that is less than Edenic.

In the early 1970s, Joe Shoemaker did not buy the idea that nature was to be found by heading out of town. He and his friends launched their boats into the South Platte, planning an excursion through the waterways of Denver. As the water received their boat hulls, a municipal dump truck backed to the river. Joe stopped the dump truck from unloading its trash into the river and devoted much of the next four decades to giving the river "back to the people." With his fellow river visionaries, he used aesthetics as a guide, focusing not on preservation of the pretty spots but on reweaving life's community in the ugliest places on the river, including what is now Confluence Park.

Denver's Greenway Foundation now continues this work. The foundation's informal motto, inscribed on office equipment and inked into some skin, is "MSH." Make Shit Happen. *Escherichia coli* is not what they have in mind, although bacterial counts have declined thanks to the foundation's work. Instead, Shit Happens through meetings with city officials, alliances with state government, fundraising for riverside projects, management of youth education and internship programs, meetings with owners of downstream water rights and upstream dams, and advocacy for and celebration of the river through public events and the media. A philosophy underlies all this work: people and the river are not separate. If people acknowledge and act on that fact, good will emerge. Beauty and ugliness are guides in this work.

Joe Shoemaker was a Republican state senator and chair of the state's Joint Budget Committee and Senate Appropriations Committee. He was a master of MSH, achieved not through solitary heroics but through working within the human social network to change the river and its lands. His insistence that parks should be accessible to all, even when they were in less attractive parts of town, was a vision of social justice. Beyond justice in the human community, he understood reciprocity in political ecology. People don't just visit the river; the river becomes part of people. In political terminology, he built an energized constituency for the river. In ecological terms, human politics is part of river dynamics, and the river is part of people's being. Strengthening connections ensures the survival and future vitality of the network, even as individual people, parks, or dams pass away. The public memorial celebration of Joe Shoemaker's life and work was held at Confluence Park in 2012. State dignitaries and Joe's friends stood on the walkway in front of the beaver-nipped cottonwood tree and added their voices to the flow of stories that is the river.

Shoemaker and the Greenway Foundation are not alone in their work. Other charitable groups, local governments, and Denver businesses now compose an ecosystem of river advocates. The river's voices are not all political movers. The engineers who puzzle over the best ways to repair old sewage and storm-water pipes, the geologists who plan water-cleaning retention pools, the biologists who manage the microbial communities of wastewater treatment plants, and the teachers who bring their students to the water are the network of quiet action that keeps the rivers alive.

The fallacy of believing that the human community dwells outside of nature is exposed at Confluence Park. City policies that emerge from primate minds affect the movements of all life-forms—people, bacteria, beavers, and cottonwoods—and meet in Cherry Creek and the South Platte. In the nineteenth century City Hall was washed away by the river. Now, from higher dry ground, local government is one part of the confluence of relationships from which the river is made.

On an August afternoon at Confluence Park, a busload of disabled children arrive with half a dozen kayaking instructors. They head to the chute of rapids and swirling pools that engineers built into the South Platte when they designed the park. One after another, the children paddle the rapids in tandem with an instructor. An African American seven-year-old uses his blade leg to jump into the boat. He leans forward as the bow raises a spray. His mouth turns from an apprehensive frown to a big O of surprise, then to an openmouthed smile of delight. Lifted from the boat by another instructor, he high-fives his kayaking mentor. The run over, the boy then turns his attention to the sands of Cherry Creek, dodging among rocks to poke for shells, a favorite activity for kids at the park. John Muir might have paused awhile and smiled, seeing, as Joe Shoemaker did, beyond the "doomed . . . toil in town shadows while the white mountains beckon."

Later in the afternoon, a Latino family sets their blankets and bagged picnic on the grass near the cottonwood. The mother supervises, positioning grandparents and youngsters, delivering food to her charges. Two girls gulp their sandwiches, then break away from the group, drawn by the river. With self-forgetting pleasure they fall to work building a sand castle on Cherry Creek's foreshore. Whispering a secret to each other, they top their turrets with a sprig of cottonwood and a small sunflower plucked from the willows.

Callery Pear

Manhattan
40°47'18.6" N, 73°58'35.7" W

One way for a stranger to crack the plate metal of Manhattan's social anonymity is to wire a tree. It is a clear-skied, chill April morning and I am standing at the intersection of Eighty-sixth Street and Broadway, using a fleck of removable wax to ease a sensor onto a pear tree's bark. The sensor, an electronic ear, is the size and color of a black bean. A blue wire threads through two book-size processors to my laptop. Headphones complete the tree-to-human sonic pathway. At one end of the wires, my ears. At the other, a street tree planted in a rectangular sidewalk opening barely larger than the tree's base. The trunk is as thick as a broad human torso, the tree's upper branches reach the third floor of the street's apartments, and outspread limbs canopy both the sidewalk and one lane of Broadway.

At the wired tree, the stream of pedestrians stutters and knots. Eyes meet as people gather and talk. Their curiosity starts with the gadgets but within a minute moves to the tree. After preliminaries— asking the tree's species name—the group of strangers talk of pleasure and of worry. The tree is gorgeous in spring when it blooms. The salt here is terrible. The branches cast such marvelous shade in the summer. The city is planting more trees, just in time. Then a measure of kinship slips into the conversation. The tree is like us, one man says; it needs the stimulus it was born into, even the crazy noise. Finally a ponytailed white man starts an oration on his 9/11 conspiracy proofs. The knot unties. Lying, I tell him I'll read his Internet screeds, and I am left on Broadway with the tree, my blue wire, and passing eyes once again studiously averted.

The bark-pressed sensor records sound vibrations in the solid parts of the tree, ignoring waves in air. Human voices tickle the upper surface of the bark, leaving a ghostly imprint in the recording. Our words tremble, then dissipate in the bark's sponge. The vibratory world of the tree—the sounds that flow through its being—is dominated by more forceful talkers. The Seventh Avenue Express hammers through a buried subway tunnel ten paces to the east of the tree, on lines two flights of stairs under the street surface. The slamming clatter of wheel on metal track, so familiar to the ears of straphangers, flows rootwise into the tree, shaking the wood a fraction of a second before the sound spurts from the street grating. Pressure waves travel ten times faster in concrete and wood than they do in air. In one second a bang, squeal, or shudder travels through air for just over the length of a city block. Inside the solid material of a road, the same sound travels more than three kilometers in a second, almost the length of Central Park. In granite curbs sound's velocity doubles again. Not only is sound faster in these hard media, energy also moves with little attenuation. As I sit on the low tree-guarding fence under the pear tree, the subway judders my buttocks and spine through the metal rail, but the airborne waves stir only the tiniest hairs in my inner ear.

These movements become part of the tree. The city dwells within the pear. When a plant is shaken, it grows more roots, investing proportionally more of its bodily resources in anchorage. Roots stiffen, making them more resistant to sway and bend. Their length-wise strength increases also, ropes adding extra strands of cellulose and lignin. A city tree therefore clings more tightly to the earth than its countryside cousins. Tree trunks respond to movement by thickening girth. Inside, the cells that compose wood grow more densely, with stouter walls. Nietzsche's maxim "From the war-school of life: What does not kill me, makes me stronger" might be revised, stepping away from individualism. From the relationship school of life: What does not kill me becomes part of me, erasing another boundary. Flexure of a tree brings within what was outside. Wood is an embodied conversation between plant life, shudder of ground, and yaw of wind.

A beer truck parks in front of the tree and I feel the throb of its diesel engine as a gut pulse and growling throat reflux. Under my palm the trunk has a gentle, barely detectable tremor. The driver slams up the metal cargo shutters, playing my skull like a wash-board. My eyes blink and my vision swims for a millisecond under the smack of corrugated reverberation.

The lower frequencies of the truck's engine pass through and around the tree's leaves unimpeded, like ocean swell passing over a sea-grass stem. Higher notes—the tight riffs of a sax busker, the brake yip of a delivery bike before it runs the red light, a woman's laugh of delight into her ear-pressed phone, the piping note of an agitated sparrow—roll outward in centimeter-long pressure waves, the size of pear leaves or smaller. Thousands of leaves that act as wax-glazed reflectors vault the sidewalk, a dome in which the upper registers of the city's sounds are cupped and retained while the bass runs away. It is a subtle change in timbre, but as I tread the concrete slabs, I hear a lightness under trees, a flick of brighter sound. In spaces between plantings, the acoustic surface loses its gilding. Sounds fly as if re-leased into a wide hall. In a few paces in Manhattan, we traverse glade and canyon. I feel this on my skin more than hear it in my ears, a wavering breeze of sound.

Like the pear tree, our whole body is sonified. Hearing is not only an aural sensation. In my ear canals hair bundles float enclosed in a few drops of ocean water. Rooted in a cell surface, each bundle turns flickers of high and low air pressure into nerve signals. The bundles translate fibrillating fluid into electric charge, thence to the brain. Vibrations arrive on many paths. Bonelets of the middle ear lever onto the eardrum. The skull's temporal bone wraps the inner ear and shudders with sound from without and within. Cranium is a dish and a drum, mouth is a wet horn, throat and spine are passageways from the lower body. Torso is pumpkin, half seedy gut, half hollow lung gourd. Skin crawls across face and ears and down the ear canal. Earrings are antennae, poking into lost frequencies. Before awareness, nerves mingle, chat, and decide what will be raised to consciousness: Hearing is modulated by tongue taste, emotion, foot soles, hairs on skin. What we perceive is the conclusion of our body conversing within a purring, stridulating world.

The city's acoustic extremes push these truths into awareness, teaching me that sensation is not stimulus and response along single axes but manifests in consciousness as a community effort. Thirty steps north of where the tree stands, a food vendor cooks meat and fixings on a sizzle plate. His food is necessarily salty and spiced. In the street din we'd taste nothing without a kick to the tongue. Only in quiet surrounds can flavorings be toned down and subtlety appear. The palate is collateral damage in Manhattan's acoustic arms race among restaurants. At tables as loud as factory assembly lines, even sweet, spice, and salt are hard to perceive, let alone the whispers of fruits or leafy greens.

Our skin too textures what we hear. In the buffeting wind of passing trucks, wakes that send tree branches lurching, our sonic interpretations may be garbled. Laboratory experiments show that what we experience in our minds as "hearing" comes partly from our ears and partly from the rest of our bodies, especially the movement of air over our skin. Silent air brushing over us modifies what our brains perceive. When air puffs against the touch receptors in our skin, we hear lip-aspirated words, even if our companions spoke only from the throat. We hear "da-da" as "pa-pa," "tar" as "bar," "dine" as "pine." For words whispered into our ears, this sensitivity to touch is perhaps not surprising. But air gusting on our skin changes what we hear even when it is our hands, not our faces, that experience surges of air. So when passing traffic stirs the sidewalk air or when a building deflects a downdraft onto pedestrians, the city's physicality merges into our perception of the social world. There is no distinct boundary between the "environs," the outside, and our innermost experience, consciousness.

Interior senses—emotion, thought, judgment—weave themselves into what seem to be exterior stimuli. Pitch and genre of music change our perception of food and wine, with bitterness emerging on our tongues from the lower registers and zing from a bouncy tune. A Tchaikovsky waltz evokes a feeling of sophistication on the palate that is absent when dining with rock music played on synthesizers. City sounds also seem louder when they emerge from what we believe are inappropriate contexts—parks instead of streets—even when the amplitude of the sound itself does not differ between these places. "Noise" emerges from the truck engine, but also from our inner narrative of what belongs and what is alien.

Human voices amid the tumult of traffic and machines become louder and higher, with lengthened vowels. We punch more energy, bellowing lungs to larger capacity and kneading facial muscles into expressive dance. We're not alone in this. Birds move their songs into higher registers amid traffic noise, skimming them over the city rumble. They too must be louder to be heard. Those species that cannot adapt lose their acoustic social network and, thus severed, disappear.

The most common nonhuman sound at the pear tree is the chatter of the starling. Its squeals and clatters dance on top of the viscous mire of road sounds, escaping to an unclaimed acoustic realm.

The sounds of the city combine with other sensory innovations to confuse many species here. The unheard shimmer of electromagnetic radiation from electronics and radio signals is stronger in the wire- and transmitter-filled city than in the countryside, and this noise disrupts birds' compasses. In a haze of radio waves, they know not where to turn. Diesel fumes bind and twist flowers' aroma chemicals, befuddling bees. Moths cannot follow scent trails through the city's perfume. The microbial inhabitants of tree leaves seem unable to find and talk to one another; their diversity is lowest in cities. Only a few species are able to make their way in this novel world. The Callery pear tree is one, thriving by wheedling its way into human affection.

At 10:00 p.m. the full moon silvers the reflective surfaces of the pear tree's highest flowers. Light takes an indirect route. Petals receive their glow from the Moon's reflection in the windows of the street canyon. The Moon is itself a glass from which the Sun gleams. Light weeps from flower to flower, descending. From below, the amber of a shop front rises to meet the moonlight, mingled with a plume of red from the newsstand's neon. Sun, to coal, to bulb, to petal. Broadway is arched by sunlight's ambling earthways. A few blocks southeast, whole side streets are tunnels of lambent pear blossoms. Manhattan receives a brushstroke from Yun Shouping's Moon-inked flower paintings.

In the morning these imaginings of seventeenth-century China flee. The beer truck parks. Ten thousand white pear blossoms shake under the jab of piston rods driven from plosive combustion chambers.

The pear owes its presence on Manhattan's streets to *Erwinia amylovora*, the botanical cousin of *Salmonella*. The bacterium is native to North America, with a taste for plants in the rose family, a taxonomic clan that includes apples, blackberries, hawthorns, and pears. When colonists brought the European pear to North America, *Erwinia* went to work on the naive arrival. Like bees in a hive, *Erwinia* cells continually exchange information with one another, using their collective knowledge to decide when to produce chemicals that lead to attacks on plants or defensive actions against bacterial competitors. In the early twentieth century, this bacterial intelligence swarmed and scorched America's orchards. Blackened leaves and stems twisted from branches, giving *Erwinia* its common name, fire blight. Crop losses approached 90 percent. In 1916 the chief of the U.S. Bureau of Plant Industry commissioned the Dutch-born botanist and explorer Frank Meyer to gather from China "as large collections as possible" of Chinese species and varieties of pear. The agronomists hoped that crossbreeding Asian species with the European pear might impart some blight resistance to American orchards. Meyer sent back to the United States sackloads of seed. Of the Callery pear, a species named for an earlier European explorer, Meyer said that its ability to thrive in all kinds of hostile soils in China was a "marvel." Meyer never saw his trees on American soil. He drowned in the Yangtze River as he was ferrying himself to another collecting spot. His sylvan legacy, though, grows over much of North America.

As the plant breeders had hoped, some variants of Callery pear proved resistant to blight, and the species is now used as a rootstock for many other pears. Among the experimental orchards a few individuals stood out, particularly in spring. These white torches of petal caught the eye of horticulturalists in the 1950s, when suburbia was spreading and in need of fast-growing, pretty trees. One individual from Nanjing, the "Bradford," named for a Maryland plant breeder, was plucked from the crowd and cloned by grafting. From this one tree millions of streets, housing estates, and industrial parks are now planted. Botanists remember the 1960s and 1970s not for tie-dyed summers of love but for the monochrome, asexually propagated Bradford.

On Mannahatta, the Lenape "island of many hills," oaks, hickories, and pines grew on what is now the intersection of Eighty-sixth Street and Broadway. A few dozen paces to the east, a stream meandered through meadows maintained by Lenape fires. I learned this ecological history from Eric Sanderson's explorations of the area's old maps and writings. He cites Dutchmen from the 1630s—Johann de Laet, David Pieterszoon de Vries, and Nicholas van Wassenaer—who wrote of the "wonderful size of the trees" on an island stocked with "great quantities of harts and hinds . . . foxes in abundance, multitudes of wolves . . . beavers in great abundance," where "birds fill the woods so that men can scarcely go through them for the whistling, the noise and the chattering." Nearly four hundred years later, in dozens of hours of observation of the botanical legacy of another Dutchman, Frank Meyer's pear, I observed not one bee among the blooms. The gnats and mosquitoes that are my companions at most other trees were absent. I saw five bird species: European starlings, Eurasian house sparrows, Eurasian rock pigeons, a red-tailed hawk threading the highest part of the canyon of buildings, and a warbler that lit for two seconds in the tree before darting down Eighty-sixth toward Riverside Park.

The diversity of nonhuman life plummeted as the coastal island changed from Mannahatta to Nieuw Amsterdam to New York. This pattern of biotic diminishment within cities was repeated across the globe. Cities, on average, are home to just 8 percent of the birds native to the surrounding countryside. Plants fare a little better, retaining one quarter of their native diversity in urban areas. Along with these losses of native diversity comes homogenization. Ninety-six percent of cities across the world are home to *Poa annua*, a low-growing grass originally native to Europe. *Poa* evolved through the hybridization of multiple other grass species. This fusion of parental lineages gave the grass many parents, a bequest of genetic memories that allowed *Poa* to adapt to cities and become a nimble follower of humanity's worldwide urban wanderings. Bird communities too are dominated by a few cosmopolitan species. The pigeons, starlings, and house sparrows that I saw at the pear tree are the same species that I would see in at least 80 percent of the world's cities.

Such patterns seem to lend credence to the antiurbanism of many environmentalists. Yet cities occupy just 3 percent of the world's surface and house half of the human population. This intensification is efficient. The average citizen of New York City releases less than one third the U.S. national average amount of carbon dioxide to the atmosphere. Unlike those of sprawling cities like Atlanta or Phoenix, New York's carbon emissions from transportation have not risen in the last thirty years. Denver, despite its profligate lawns, waters one quarter of Colorado's population with 2 percent of the state's water supply. Therefore, the high biodiversity of the countryside exists only because of the city. If all of the world's urban dwellers were to move to the country, native birds and plants would not fare well. Forests would fall, streams would become silted, and carbon dioxide concentrations would spike. This is no thought experiment. These outcomes are manifest in the cleared forests and extra carbon dioxide emissions that result from several decades of city dwellers' flight to suburban and exurban peripheries. Instead of lamenting a worldwide pattern of biological diminishment in urban areas, we might view statistics on bird and plant diversity as signs of augmented rural biological diversity, made possible by the compact city.

Even within the urban center, where buildings and roads occupy 80 percent of the land surface, other species can survive and, sometimes, thrive. Two beavers, symbol of the Dutch fur traders, crouch to this day on the flag of New York City. For two hundred years the pair fluttered over friendless rivers. Now, as in Denver, the animals have returned, attracted by the cleaner water and fresh vegetation of the Bronx River. A few blocks east of the pear tree, in Central Park during the spring migration of birds, I saw thirty-one species of bird in as many minutes, most of them in plantings of native vegetation. Some of these species were residents; others were migrants making use of the park's greenery as they moved along the coastal flyway, headed north toward fir trees in the boreal forest. "Whistling, the noise and the chattering" has not entirely disappeared from Mannahatta's forests.

Thanks to previous generations of planners, trees canopy 20 percent of New York City's land surface. Human hands planted almost every one of these trees. In 1904 Broadway was excavated, then refilled to build the Eighty-sixth Street subway station, one of the city's first twenty-eight stops. Only one tree survived the digging and it stood close to where the pear tree now grows. Arthur Hosking's street photographs from 1920 are too indistinct to precisely locate this forebear, but the black-and-white images show a streetscape with only a few short saplings in Broadway's median and sidewalks almost devoid of plant life, save for one or two trees on some blocks. Through the following decades, extensive plantings greened the city, but in the last thirty years, vegetative cover has dwindled. Buildings took over green spaces and relatively few trees were planted. The MillionTreesNYC project, launched in 2007 by the city and the New York Restoration Project, attempts to reverse the decline through planting and caring for at least one million saplings. By the winter of 2015, the goal of planting a million trees had been met, but the longer-term aim of reducing the overall loss of green space remains elusive.

Bradford pears are not part of this million-strong flock of trees. The species has fallen from favor, at least among professional horticulturalists. A genetic quirk of the original tree from which the Bradford was cloned endowed all the descendants with weak limb attachments. Trees shatter under the weight of ice or snow, even trees strengthened by the rumble of buried subway lines. The Bradford therefore requires more time from arborists than do most other trees, both to repair damage and to shape the healthy limbs into stronger forms with few weak, acute-angled branch crotches. The sylvan star of the 1960s turned out to be a high-maintenance embarrassment for future generations. Being from another continent also counts against the tree in our present-day calculus of ecological value. Native trees are richer communities, hosting many more leaf-munching and flower-supping insect species. This diversity feeds predators—spiders, wasps, birds—adding yet more variety. Callery pear is shielded by chemical defenses against which local animals have yet to evolve countermeasures. The pear's leaves are therefore tidy and whole, unchewed by caterpillars or leaf miners. Formerly we saw this as a decorative virtue; now it is an ecological vice. The Bradford's behavior outside the city further sullies its reputation. The tree cannot self-pollinate, but when its pollen or ovules join with those of other Callery variants, the pebbly fruits bear fertile seed. One hundred years after Frank Meyer shipped his pear seeds from China, the government that sent him classifies Callery pear as a weedy invasive.

The Callery pear is not alone in having changed its position in the hierarchy of human esteem. Hedge-forming European and Asian privets were planted throughout American gardens of the eighteenth and nineteenth centuries, guided by the approving words of government botanists and private nurseries. These imported privets now cover millions of acres of American forest. Most modern ecologists and horticulturalists regard privet as a noxious weed. Toadflax, a pretty yellow-flowered herb, came to the Americas as a medicinal and ornamental plant from Europe. The species shoulders its way into riversides, meadows, and fields across the continent, sometimes growing in patches that cover thousands of acres. Hundreds of other species are in the same category. They were once lauded and imported; now they are condemned. We celebrate what our forebears disdained, the provincial, the local. We suppress the immigrant, the species that we judge not to belong. But this judgment emerges from a pragmatic and changeable place. We no longer need Callery pear for its blight resistance, privet for its hedges, or hundreds of other immigrant plants for their medicinal qualities. Should blight infest all American orchards, or metal fencing become too scarce, or the pharmacy close its doors, our stories about which species belong would undoubtedly change. The human mind is a wild and changeable creature, finding its place within life's networks and reaching out to modify them as our needs evolve.

Late at night traffic eases. The sidewalk crowds thin as people lock the doors to the high-rent apartments that frame the Upper West Side's streets. The pear tree now hosts those who cannot sleep and those who could but, for want of a bed, must stay on the move. Perched on the tree guardrail, head tipped forward, a woman bundled in a dirt-smeared overcoat coughs. Hers is not the sunny back-of-the-mouth cough of the afternoon's kids, who laughed in the puffs of dust that shoot out of the subway gratings as trains pass. The woman's cough comes from a crack-lipped mouth in a lined face. Between drags on her cigar stub, her humped shoulders shake as fluid convulses inside ragged lungs. The sound latches on to some part of my nervous system and evokes a tremor of dread. Somehow human ears and brains understand the meaning of pulmonary distress.

The cigar's blue efflux arrives in lungs that the city's air has already permeated and damaged. Every inhalation is downwind of the tailpipes of New York's City's nearly two million vehicles and the chimneys that vent furnaces burning a billion gallons of heating oil every winter. Some of the oldest and most prestigious buildings on the Upper West Side and in other higher-rent areas of the city warm their inhabitants by burning tarry sludge, a low-grade fuel oil. The smokestacks from these fires rival Auld Reekie's nineteenth-century chimneys. In the last decade the dirtiest oils have been phased out, cutting soot by one quarter and acid-forming sulfur dioxide by nearly three quarters. Even now, though, when New York City's air is the cleanest it has been in fifty years, the Upper West Side stands out as a hot spot on air-pollution maps.

Later the coughing woman moves on and the rain starts. Under the tree, I feel no wetness for several minutes as water adheres to leaves and starts its trickling passage down twigs to the trunk. Unlike with Amazonian rain, I hear only a few clicks from leaves. All other sound is road-killed by the hiss of water crushed and flung under spinning rubber. Even the thunder that accompanies the rain is overwashed by tire sounds until a crash comes directly overhead. With ears shut down, skin once again becomes a rain sensor. Only when the canopy becomes saturated does my face start to feel the cold sting of pear leaf water drops. The open-skied sidewalk three paces away receives far more water than I. By capturing water on their surfaces, leaves reroute flow. Some water adheres to the tree surface and never reaches ground. Most of the tree-captured rain finds its way to the bark, where it descends to the soil at the pear tree's base. Here the water soaks to porous ground, rather than running on the impervious street and thence to gutter drains.

Trees' interception and diversion of rain has downstream benefits. More than half of New York City's street runoff flows into pipes that also serve as conduits for sewage. In a heavy rain, sewage treatment facilities can be overwhelmed and burp untreated sewage into streams and rivers. By buffering storm-water flow, trees reduce the number of river-fouling "combined sewer overflows." Trees and the soil around them, along with new retention ponds, have lowered the proportion of storms that cause sewage to flow to rivers from 70 percent in 1980 to 20 percent today. Fish in the Hudson River are therefore partly sustained by the soils and trees of Broadway.

I lean my hand against the tree's bark and feel a slight yield under the pressure of my palm, a cold ooze. The crinkled fissures in the bark are runnels and drip points, like braided rivers miniaturized and turned vertical. Bubbles skim the waterways. I pull away and the skin of my withdrawn hand shocks me with a stain of liquid smoke. By resting against the tree, I've palmed a slurry of ash. I hold out the hand and, over several minutes, the rain spats and sluices away the grime. At the base of the tree, foam accumulates over pools of water as dark as salt marsh mud. The gutter too runs black, clouded by the city's exfoliations. This tree has intercepted more than water. As rain continues, a patchy sheen of emerald emerges on the bark. Its surface washed, the algal community can once again emerge into the storefront glow.

Particulate pollution, leftover grains and specks from combusted fuels, settles on bark and leaves. When rain comes, the encrustations drain to the ground. On dry, gusty days, the accumulated detritus can sweep once more to air, like a vacuum cleaner bag shaken out. The overall effect, though, is cleansing. In summer leaves augment the effect as they draw soot and pollutant molecules into pores. In the wet interior of the leaf, these chemicals dissolve and are incorporated into the plants' cells.

Not all trees can survive a poison-swallowing exercise, but the rigors of Callery pear's life in the less hospitable soils of China prepared it for the city. Inside the tree's cells, chemicals bind and render harmless metals such as cadmium, copper, sodium, and mercury. When geneticists engineer the DNA responsible for producing these chemicals into bacteria, the resulting cells can detoxify soups of metallic poisons. If such innovations could be applied outside of the laboratory, the Callery pear might, through the genes hidden inside its cells, help to clean up industrial waste. Inside the tree, chemistry not only helps the tree survive the city's aerial pollution but also gives it an edge when winter de-icing salt soaks down to the roots.

When doormen from the neighboring apartment buildings and trucks on Broadway scatter salt in winter, the pear can endure in a way that more tender species, such as maples, cannot. Like cottonwood's experience in the western United States, many generations of experience in Chinese soils has allowed the pear to thrive downstream of human attempts to melt ice on pavement.

Combined, New York City's five million trees yearly remove an estimated two thousand tons of air pollutants, in addition to more than forty thousand tons of carbon dioxide. In optimal summer conditions, in areas with a thick canopy, trees can remove 10 percent of some pollutants every hour. But conditions are seldom so good and the supply of new pollution is ample. Over an entire year, trees remove about 0.5 percent of the city's air pollutants. This average hides the disparities. People in well-planted neighborhoods breathe easier than those who live where trees are sparse or absent. Depending on where she lived during her 1960s childhood, the woman's rattling, wet cough may have been born from cigars, exhaust pipes, and treeless sidewalks.

The chambers of human lungs and the interior spaces of tree leaves are biological soot catchers, linked to one another and the city. Out-of-town landfills concentrate and sequester large waste, but we all bathe in the garbage cloud of the sky, home to billions of tiny pieces of discard. The city's tree-planting program now uses these connections to plan its work. Where maps of asthma hospitalization rates coincide with maps of treelessness, the Department of Parks and Recreation revegetates whole blocks. This tactic contrasts with the former method of planting in areas where residents had called the city to request trees. In general, this made leafy areas leafier and neglected places in which trees had disappeared from streets and therefore from the imagination. The city now uses both approaches but has lately taken on many more blockwide plantings, establishing an urban forest canopy where people need it most. One does not require a sulfur dioxide monitor to know the results. Just as the sonic texture of a street changes as we walk under trees, then into the open, so too does the taste of air. A well-planted block has a slight mouth tang of salad and earth. Molecules drift from pores in tree leaves and out of the soil around the bases of tree trunks. We sniff and tongue a forest as we breathe. Between planted blocks, away from trees, the air tastes of thin brown acid, the aromas of engines, gutter grime, and asphalt. The contrast is most striking when walking from a bus-jammed thoroughfare into a park. The wooded glade or open lawn presses distillate of leaf to our mouths.

Tree planting is the first necessary step to establishing an urban forest, but even the most lovingly loosened soil around a freshly planted sapling is insufficient to ensure the tree's survival. Young trees are killed by traffic accidents, vandalism, drought, pollution, poisoning by excessive dog feces, soil depletion from sidewalk washing, and the thousand other slings and arrows of urban fortune. In industrial areas and near vacant lots, 40 percent of plantings will not survive to their tenth birthdays. Tree species differ in their hardiness—Callery pear is the most resilient of all street trees in New York, with a 30 percent better survival rate than the least hardy species, the London plane tree—but a tree's relationships with the human community override these botanical differences. Survival increases when saplings are embedded within the human social network. A tree planted by its human neighbors will live longer than one placed by an anonymous contractor. When a tree bears a tag naming it and listing its needs—water, mulch, loose soil, no litter—its probability of survival jumps to nearly 100 percent. A street tree that is granted personhood and membership, one that is noticed, loved, and given identity and history, lives longer than a municipal object, arriving with no context and living with no collaborators.

Psychological bonds to trees in cities are often fierce. In my experience of talking with people about trees, New Yorkers echo, in their own way, Amazonian Waorani. Relationships with trees are deep and personal. Outrage flares when the conversation turns to trees heedlessly damaged, whether the tree is a Manhattan street tree damaged by building construction or an Amazonian ceibo tree cut to make room for an oil-prospecting road. These wounds are keenly felt by the people who live with the tree, people whose lives are wrapped into the tree's own narrative. In New York City and in rural areas alike, a conversation about the future of trees will animate and connect strangers. Trees, especially those that grow close to human homes, are portals to unselfed experience: the Others that stand in front of the New York apartment, evoking old memories of forest through the leaves' susurration and their springtime flames of green. When such portals are scarce, their value increases, bringing city dwellers closer to the knowledge gained by those who live in forests and orchards and who are daily reminded of the centrality of trees for human survival and flourishing.

Concern for sylvan doorways to beauty is not universal. The diversity of human relationships with trees is evident in the state of the soil around tree trunks. The pear tree has a knee-high metal railing, provided, as are most tree guards, by the owners of the apartment building that backs the sidewalk. Plantings of *Coleus* and *Begonia* sometimes necklace the trunk's base, massing the sidewalk opening with annual blooms. Both plants came originally from East Asia, granting the pear an unintended biogeographic homecoming. For most of the year, the soil is a bare canvas onto which the city flicks its pointillist ornaments. On a day in midsummer, I counted half a dozen cigarette butts, nine pieces of bubble gum (with two more pressed into bark crevices), a grape juice can with a jaunty straw, a broken rubber band, a wadded newspaper page, and one blue plastic bottle cap.

The block to the south of the pear has no apartment staff to tend trees. Instead, the grocery store daily guns the sidewalk with water jets, scouring concrete and leaving the sidewalk's ginkgo tree with its upper roots flailing in a soilless cavity. To the northeast someone has constructed a tree guard from the wire sides of disassembled milk crates. Hand-lettered signs with thrice-underscored *DO NOT* exhort dog owners to take their business elsewhere. Another resident has invested in a hardware store fence and a carpet of marble chips. Five trees north of the pear, food vendors park their carts next to unguarded trees and the footfalls of many hungry people press soil to the same hardness as the sidewalk. The scaffolding associated with nearby apartment renovation on Eighty-sixth Street cut light and water to trees for two years. Come December, the same weakened trees are trussed in lights, powered by electrical sockets whose supply runs under the sidewalk to the tree. Several blocks east, close to the museums and Central Park, soil is hand fluffed and, depending on the season, crowned with ruler-straight purple tulips or a swirl of Maine fir boughs. If trees are portals to the beyond-human community, tree guards and planting holes are the windows through which we gaze back at the diversity of people.

Tree guards are designed to protect roots and trunks from cars and pedestrian trampling. But people too sometimes need protection from trees. If left unpruned, trees shed limbs from above, sometimes with disastrous results for those standing below. In the second year of my time visiting the tree, several branches grew stunted leaves after the springtime bloom. The leaves and their branches crusted and clenched to brown. This Callery pear was not immune to fire blight; the bacterium entered through the flowers, then invaded twigs and leaves. The doormen and janitors gathered at the entrance to the adjacent apartment building told me how worried they were. Loss of street trees on this block would sadden them, they said. Dead tree limbs could also clobber a passerby. Later in the year a pruning crew lopped the branches, cutting each at the collar of its trunk attachment. This was expert work, with cuts at just the right position and angle to allow healing.

Despite the dangers of falling branches, in care of dead and dying tree limbs New York City has been inconsistent over the years. In 2010 the city's tree-pruning budget was itself pruned. The following year the number of tree-injury lawsuits filed against the city soared, resulting in several multimillion-dollar settlements. One settlement, for a branch that caused severe injuries to someone sitting on a bench, cost the city more than two times the amount of its previous annual pruning budget. When a tree falls in New York City, not only will it be heard, but someone with good legal counsel is likely to be under it. In 2013 funding for tree care was restored. People walking under the pear tree on Eighty-sixth and Broadway are the beneficiaries of this action. Sickened limbs are now cut away. In urban forests, more than any others, the benefits that we derive from our woody cousins must be reciprocated with attention to every branch.

Morning rush hour comes and the sidewalk turns to a river of feet, umbrellas, and shoulders. The river's sounds are shoe soles slapping, riding-crop snap of men's leather-bottomed business shoes, *tok-tok* of the fashion teeter, chuff of runner, dog toenail tick, and the scrape of tired amblers. The orifice of the Eighty-sixth Street subway station swallows us all, like a river-bottom sinkhole, then it bubbles more to the surface. Crosstown buses roar from their stops, leaving wakes of commuters whose swell is soon dissipated into the main sidewalk channels. At the Eighty-sixth Street and Broadway intersection, two currents meet, a turbulent merger whose rhythms are set by the red and green sluice gates of traffic lights. The tree stands amid this motion, firmly lodged close to the riverbank. Its immovability creates a pool beside the run and riffle of pedestrian flow. With a fin flick, people escape the torrent. They drift in the calm created by the physical obstruction of the tree and its metal rail. In winter a moment's glance at the ground reveals the pattern. Under the tree, snow is broken only by a petal fall of undirected footsteps. Immediately adjacent the sidewalk is pummeled to slush by footfalls, all of them moving along the same axis. Like the cottonwood in Denver that creates habitat on the riverbank, this pear has opened new possibilities for humans in its surroundings.

The still space around the pear tree seems to be a gendered and raced space. Of the dozens of people I saw step out of the march on the sidewalk, three quarters were women, comprising many races and classes. Of the men, none were white, unless I count my own presence. There was no shortage of white men running the rapids in front of the tree. Inside the quiet water next to the tree, people talked on their phones, lit and savored cigarettes, rearranged bags or umbrellas, stood in repose, or unfolded the newspaper while perched on the railing.

In New York, to obstruct pedestrian traffic without a "legitimate purpose" or, worse, "with intent to cause public inconvenience," is a disorderly conduct violation of the state's penal code. The penalty for such transgression may include a fifteen-day stay inside the state's Rikers Island facilities, although most people get off with a fine and community service. Of course, in a city, to walk or stand on a sidewalk is to obstruct someone, and so the law can serve as one strand of a net where police may stop anyone at any time. Street trees are in perpetual violation: obstructionists all, maintaining, as Howard Nemerov observed, "comprehensive silence" about intent and purpose. Like the trees, contemplative humans are scofflaws. To attend without purpose is disorderly. To stop moving is a violation. Standing under a city tree is an act of microsubversion, a fact perhaps understood by the well-tailored men who so seldom stepped into the pear's shade. It is not just wood that takes into its being the thrum and rule of the city.

When I step out from the tree's space, my presence inadvertently changes from microsubversion to microaggression. A white male standing on the sidewalk creates another form of gendered space.

With my back against a storefront, notebook in hand, I was a stake pounded into the edge of the river, blocking perhaps 10 percent of the sidewalk's usable width. Within a few minutes of my taking this position, between one and five men would join me, standing a meter away, eating their street food or talking on their phones. My fellow standers were never women; the men were usually nonwhites. After three repeats of this experience—two to awaken and one to experiment—I saw that I was creating an obnoxious bubble that took far more than 10 percent, inconveniencing people on the public walkway. My stationary bulk was a passive version of "manslamming," the act of men not yielding to women on the sidewalk, to the point of bodychecking. When labor organizer Beth Breslaw experimented with walking like a man, every walk became an ice hockey–like traverse of New York City's streets. As she walked, almost all males refused to yield even a fraction of "their" space. Chastened by what I had learned by standing against the wall, I moved my note-taking position to the nooks created by doorways and subway exits. As long as I retained the notebook in hand, people paid me no heed. Standing with a vacant countenance, though, caused another bubble to grow, even when I was out of the flow. Under the tree I could sit or stand without my presence intruding into other people's spaces. The tree's pool seems to provide a partial escape from the usual rules of the sidewalk, at least in this more affluent part of town. Had I stood in a different body in a different place, the pool would surely have offered no such safety.

Street trees diversify the kinetics of human movement, especially in a city that has no "million benches" project and where the forward stride is often the only possible move. In New York City trees add meanders and bayous to what are otherwise channelized chutes. In a crowded city full of human diversity, power asymmetries, and contested space, street trees are, whether "with intent" or not, sociocultural actors. Of learning to move through the unspoken rules of New

York street herds, writer Jane Borden reflects, "New York offers only a prepositional life. No action exists without a modifier." Trees change the grammar of a street, allowing sentences that are otherwise silenced.

Trees also diversify human experience by changing the weather, in small ways on the sidewalk and at the much larger scale of the whole city. On an afternoon in late July, I rest a thermometer on the pavement under the tree: eighty degrees Fahrenheit, twenty-seven degrees Celsius. A few paces away, where no leaves cast their shade, the surface temperature is ninety-six degrees Fahrenheit, thirty-six degrees Celsius. The pear tree's obstructionist physicality creates new spaces on a vertical axis of light as well as the horizontal plane of human movement. Sidewalk vendors know this well. The pear tree shades both a cubicle newsstand and a tabletop children's bookstore. Within a minute's walk, three other vendors stand in the full sun, their parasols unable to match the reach and depth of the pear tree's canopy. The newsstand vendors affirm the thermometer's testimony: in Manhattan's summer, shade is welcome.

Stanley Bethea, owner of the book display, escapes the worst of the summer heat by working in an out-of-town summer camp, but for the rest of the year the tree's protection blunts some of the rigors of his eight-hour days on the open sidewalk. But unlike the neighboring newsstand vendors, Mr. Bethea is more concerned with floral aesthetics than shade. For him, the emergence and demise of the pear's flowers are a source of delight, then sadness. Leaves might umbrella the sun on a hot day, but their emergence in April marks the end of the tree's flowering. Mr. Bethea tells me it is hard not to resent the pear leaves for what they take away. His love for the seasons of the city's flowers gives him a botanical calendar. He'll travel many blocks to catch a tree in bloom. He knows when to expect the emergence of each of the flowers on the Broadway Mall, the wide central median, and the photograph on his business card features a flower from the mall with Mr. Bethea standing to the side, smiling. The Department of Parks and Recreation and the Broadway Mall Association, a nonprofit community group that plans and tends plantings, have nurtured threads of beauty through the city. Along these green lifelines, sound, smell, and movement become hospitable, the burning pavement is soothed, and time passes by the measure of the seasons.

I return on the same July day after nightfall, reading the thermometer by light reflected from the pearly undersides of pear leaves. Under the tree the sidewalk temperature has dropped by only a few degrees to seventy-seven degrees Fahrenheit, twenty-five degrees Celsius. In the open the temperature is a little higher than under the pear, eighty degrees Fahrenheit, twenty-seven degrees Celsius, but much cooler than its afternoon high. Concrete and asphalt have reemitted their warmth like electric heaters cranking their power into a room. New York City in July, though, has no need of radiant heaters. The unshaded rocklike surfaces of roads and buildings warm an already-sweating city. The result is an "urban heat island" where city temperatures exceed those of the surroundings by several degrees. New York City is, on average, seven degrees Fahrenheit, four degrees Celsius, hotter than its surroundings during the summer. Trees reduce this effect by blocking heat before it reaches the ground and by cooling the air as water evaporates from their leaves. Like wet handkerchiefs held over a burning forehead, they ease the city. The aerodynamic roughness of trees, the textured upper layers of the canopy, also contributes to urban heat. If a city is ringed by forest but has few trees within, roughness is high around the city and low within. Under these conditions, the swirling motion of convection cannot carry heat upward to the higher atmosphere. This stalled convection traps heat over the city. Cities in eastern North America such as Boston, Philadelphia, and Atlanta spend most of the summer under such heat lamps. Trees in cities therefore palliate fever in many ways. In New York City they save eleven million dollars in cooling costs every summer.

For this visitor from a small rural town, Manhattan seems at first a place of vast human loneliness. To catch an eye and nod hello or to offer a passing "Good morning" often causes the recipients of these civilities to quicken their steps. The informal social bonds that apply elsewhere, and likely applied for most of human history, are severed. We're then freed or condemned, depending on our temperament, to be individuals. For outsiders this is the allure and the sadness of the city. Trees, their biological communities likewise decimated, seem embodiments and allegories of the city's atomized aloneness. But over three years of visits to the pear tree, I learned that these first impressions might be true for a visitor and even for some residents, but for the city as a whole they are falsehoods.

My harmless sidewalk curiosity—a tree with earphones—was a first unwitting experiment in hearing through the wall of social silence. The small crowd was eager, garrulous, and free with their stories and opinions. A node of intense social exchange created itself ex nihilo, as if a creationist god touched the sidewalk. But gods need material to work with. Social chatter was latent in the passing silence. Then, *snap*, a man started his unhinged rants, and the nexus dissolved as fast as it was born. The social life that came into being under the tree was strong and open, but it also had a sensitive immune system: connect to others only when the signals are right; otherwise withdraw. These are the same rules that govern the conversations among tree roots, bacteria, and fungi. In the biological crowds of the soil, only selective connections are wise. Once formed, though, the links are powerful.

Sideshows may draw people together, but they do not generate persistent bonds. As I returned over weeks and then years, I saw and heard these other, longer-term connections. The people of that particular patch of sidewalk started to greet me. A few handshakes, many a "how-ya-doin'." These links have a fine spatial texture. On the northwest side of the intersection, where the pear tree was located, I became known, but I was a stranger on the southeast corner, just five pigeon wingbeats away. My schedule was irregular, departing for months, then returning to keep night vigils or to stand for hours scattered through the day. Such spotty attendance kept my connections to people loose at best.

Multiple communities exist in one place, structured by time of day, season, and month: 7:30 a.m. commuters; 2:30 p.m. kid strollers; midnight coughers and cigarette scroungers; Saturday-morning synagogue goers and Saturday-night drinkers; summer-dawn joggers and winter-afternoon dog walkers. Within each layer of this social soil, people are twined, and not just within their class and economic stratum. As on market day in my rural home, people run into one another, greet, gossip, share a glance, pull away under the tree to deepen the talk in lowered voices, laugh, tear, hug, and then walk on. But amid thousands of passersby, these manifestations of social networks are harder to detect, obscured by surrounding movement.

What seemed at first to be anonymity was in fact the coexistence of thousands of communities. I mistook the noise of the streets for the clatter of undirected atoms. What I heard was the simultaneous sounding of thousands of taut strands of relationship. Market day in the village may be abuzz, but the babble emerges from a limited number of networks. Barriers to participation are also higher. The market is missing the voices of disabled and impoverished people living with broken cars, miles down dirt roads, with only darkling thrushes for company. The countryside hides the poor and the wounded in a way that the city cannot.

Across the world, as the size of human settlements increases, the richness of the human social network rises in a pattern that mathematicians call "superlinear." A doubling of a city's human population more than doubles the number of connections among people. Not only does the number of connections increase, but the time spent communicating with others soars. This connectivity may be increasing. Archived film of people in public spaces in New York City, Philadelphia, and Boston shows that, compared with 1980, people now spend more time together in public spaces and more women are present in those spaces. Mobile phones had little effect on these connections and were used mostly when people were alone, further increasing connectivity. Cities therefore draw more and more of our being into relationship. Other measures of intensified human connectivity, and subsequent interaction, creativity, and action, also accelerate with population. These per capita measures span many areas: wages, number of jobs in research and creative fields, number of patents, incidence of violent crime, and rates of infectious disease all increase as city sizes rise. As in the rain forest, the vigor of both cooperation and conflict increases within the complexities of a city. This social change contrasts with patterns of physical growth. Instead of accelerating with increased population, the bulkiness of infrastructure decelerates. As a city increases in size, its use of land area in proportion to population size decreases. The same efficiency is evident in other physical characteristics of a city, such as the length of road and utility pipe networks or the area of land that is impervious to rain. These physical and social trends run in different directions, social connections accelerating and physical demands decelerating with increasing size of cities, but they are part of the same story. A compact and tangled environment increases opportunities for human connection. Like the shaken wood in the pear tree or the health of our lungs, the social texture of our lives is a direct product of the structure of home.

The artifice of the city brings us close to our nature. We're animals bonded, clustered around what we have made, listening to one another's songs. Other species dwell within the city, receiving and returning life through their relationships with people. Statistical analyses of our social networks in cities have yet to include the non-human. Yet the sight of moonlight on pear petals, the timing of blooms in a street median, or warblers migrating through a park are as much a part of the social life of the city as is the richness of human relationship. The fierceness of New Yorkers' attachment to street trees speaks of the vitality of the tree/human relationship. This energy is born of the city's power to knit and join, an energy that exists because of, not despite, Muir's "crowded towns mildewed and dwarfed."

Olive

Three cats—a tabby, a ginger, and a large black tomcat—have reclaimed their space under the olive tree, yowling as they bat one another and roll on the well-trampled soil. Friday prayers at the Al-Aqsa mosque ended hours ago, so the crowd striding out of the Damascus Gate in Jerusalem's old city wall has thinned from thousands to a few dozen. Street merchants shout from the plaza below the tree's low-walled enclosure, their boxes of shoes, cucumbers, mulberries, belts, plums, and coffee-making machines covering every flagstone around the main walkways. With no crowd to cloth-muffle and outshout the vendors, the cries echo from the gate's high stone facade: "*Ashara, ashara*, ten shekels!" A few soldiers stand at the plaza's perimeter, fingers asleep on gun barrels, but the dozens of black-clad security forces who stood guard this afternoon have left, their armored vans and visored horses driven back to barracks. Heat and dust are soothed by a westerly breeze that woke as the sun approached the horizon. Olive branches answer the wind with a straw-broom swish. The tree's gnarly trunk stands two meters tall, then breaks into four branches that rise another two to three meters. The dome-shaped, shaggy canopy splays eight meters across. The olive tree sits in the inner curve of the broad stone stairway leading from the road to the plaza. When the sun is high, the leaves of this tree cast the only shade here. The three cats roll on their backs, rubbing fur into the foot-crushed sage and mint spilled by an herb seller.

The black cat's ease snaps. In one motion he rights himself, then belly-slinks away from the tree and over the low wall that encloses the olive tree. I hear and see no change in our surroundings and assume that he caught sight of an unwary sparrow. Then two boys tumble down the steps that lead from the bus-choked street, scramble around a police barrier fence, and run straight for the tabby and ginger. Claws skating on the smooth surface of the stone wall, the two cats hurtle after their sharp-eyed companion—he'd perhaps seen and recognized the boys—into the lower plaza. The boys whoop and continue their run, dodging between the shopping bags and walking canes of their elders as they dart into the city's gate.

Once inside the gate, the boys run across cobbles and flagstones, heading downhill into the Muslim Quarter of the Old City. The cats took a different, lower route, descending into the Roman subconscious of the city, streets and streams running unseen, a story or more below the Jerusalem that humans now tread. Iron-barred entryways are no barrier for felines; chinks in old stone walls are portals. The cats take refuge among the city's stone memories.

Although the Damascus Gate seems old—a crenelated fortification built in 1537, in the time of Süleyman I, sultan of the Ottoman Empire—it vaults and covers spaces from the start of the first millennium. Until the 1930s these older remains were known only from sixth-century maps. Then, when British excavators cleared a plaza in front of the gate, they uncovered a buried Roman facade. Israelis renovated the plaza in the late 1970s, first sending archaeologists to explore the site. Their work revealed Roman gates, watchtowers, and roads, all buried under the modern city. In one of the newly discovered rooms, a seventh-century olive oil press stood, built by Arabic or Byzantine merchants from stone partly salvaged from remnants of Roman columns. After the archaeologists finished their dig, city planners rebuilt the plaza in front of the gate, a process that concluded in 1984, when olive trees were transplanted into low-walled enclosures at the upper entrances. The tree that the cats fled was one of these, perhaps thirty years old when it was moved, now over sixty. Today a few of the better-preserved Roman remnants are open to public visitation; others can be found by probing loose iron gates and squeezing through, but most are too tumbled, unexcavated, or walled for human access. In all these spaces, though, cats paw the tunnels. The stench of their scat and the shriek of their fights accompanied all my underground explorations.

Year-round the air under Jerusalem is dank. In summer this is hard to believe. I sit under the olive tree outside the gate, sun hammered and parched. For weeks the piss of early-rising street vendors is the only moisture that the tree has tasted. Unlike generous contributions to trees from the abundant dogs of Manhattan, these streams are too meager to damage tree roots by oversalting the soil. When rains come later in the year, the olive will likely benefit from the extra nitrogen in the urea and decomposed plant material from the market stalls. For now, the surface of the soil is tamped dust.

Olive is well adapted to such Mediterranean summer duress. The tree survives by clamping shut the breathing pores in its heavily waxed leaves and so falling into a stupor during the hottest times. As the summer days wear on, the leaves of the olive tree at the gate change their shapes, escaping the desiccating effect of the sun by rolling, tubelike, around their midribs and angling toward the stem. This motion protects the leaves' silvery, porous leaf undersides from the sun. The silver on these surfaces comes from the glint of thousands of transparent cells that stand on stalks just above the leaf surface, like microscopic parasols. These coverings hold water vapor close to the leaf surface around the pores, creating a thin layer of humid air, allowing breathing pores to stay open longer than they would in an unprotected surface.

The roots of most olive trees fan through the upper layers of soil, ready to sip on rains that are too short-lived to soak deep. But if soil and moisture should come in another pattern, the architecture of olive roots will change to fit their context. In irrigated orchards, roots cluster at the irrigation pipes and seldom burrow deeper than a meter. In loose, dry soil, half a dozen thick roots pursue divergent paths to depths of six meters. Roots of all tree species are adaptable, but olive roots are exceptionally so. The roots' dynamism is visible in the trunk. Trunks of old trees have a fluted appearance, with muscular vertical ridges separated by deep fissures. Each ridge is the manifestation of a major root. If that root finds water, the trunk and branches to which it is connected expand over many decades. Should the root die or the distribution of water around the tree change, the associated aboveground parts die back. Every olive tree older than a few decades is a composite of several of these partly independent root-to-branch segments. The tree at the Damascus Gate has two dominant trunk segments and two smaller ridges that twist against each other as they rise. In the very oldest trees, those that live for centuries and perhaps millennia, original trunks are usually absent and the tree is hollow. What we see today on these trees are trunk accretions and root sprouts that superseded older growth. The great longevity of olive trees emerges from their ability to renew as local conditions change. All this flexibility has a price. Olive trees will grow only where sunshine is plentiful. In shaded woods or cloudy climates, the tree withers for want of energy.

I follow the cats underground, dozens of meters below the olive tree. There, beyond the reach of the tree's roots, between paving stones etched with the game boards of Roman soldiers, I hear water riffling through channels cut into stone. Sometimes these conduits are crude troughs; more often they are cleanly chiseled sluices. The water arrives from outside the city's walls and drains toward cisterns buried near markets and temples. The olive tree grows almost directly above the flow that feeds the aqueducts and pools around the Temple Mount. This buried canal is just one of many conduits and storage pools. For dozens of kilometers around Jerusalem, catchments and pipes lead inward. Many are Roman or are built on what the Romans started. A few date from before the Romans. Every subsequent ruling authority has used and modified the old flows; some have added their own. Water has preoccupied every dynasty and age here. Jerusalem has grown like an olive tree; the visible city, time-worn and picturesque, is supplied by a hidden system of water vessels that must be continually revised, often at great cost, to ensure the city's survival.

According to first-century Romano-Jewish scholar Flavius Josephus, Pontius Pilate, then the governor of Judea, used public monies for the "construction of an aqueduct to bring water into Jerusalem, intercepting the source of the stream." Pilate misjudged the politics of water, "and tens of thousands of men assembled and cried out against him, bidding him relinquish his promotion of such designs." In the riots that followed, Pilate's soldiers "inflicted much harder blows than Pilate had ordered, punishing alike both those who were rioting and those who were not." Ever since Pilate, and probably in the unrecorded history before his time, ruling Jerusalem has meant paying attention to the politics of trickles and seeps.

Jerusalem's street cats have lived in a city controlled by, depending on the century, Byzantine, Caliphate, crusader, Mamluk, Ottoman, Jordanian, British, Israeli, and half a dozen other authorities. Over these millennia the cats have been chased belowground by the scuffles and cries of many political/religious revolutions, riots, and slaughters. Safe under the flagstones, their paws are wetted by Jerusalem's most persistent political actor.

The day before I followed the cats under the city, the olive tree was hung with medical equipment and fluorescent safety vests. Palestinian medics used the tree as a staging area to prepare for the coming May 15 Nakba Day—"catastrophe"—protests. These annual demonstrations sometimes turn violent and this year, 2014, tensions were high amid expanding Israeli settlements, ongoing sporadic violence from all sides, and impasses among the Israeli, West Bank, and Gaza governments. So, expecting trouble, journalists lingered in the tree's shade, with helmets and gas masks clipped to the straps of their video equipment. Inside the Damascus Gate and on the western edge of the plaza, sixty men in riot gear stood waiting, most with guns, a few with shoulder-wrapped bandoliers loaded with gas canisters or rubber-coated steel bullets. The day was blazing; their uniforms were not airy. A stack of plastic water bottles stood at their rear.

Children were the first protesters, holding keys and handmade signs. The key: symbol of the "return," when Palestinians might reclaim the houses and villages that were taken when the State of Israel came into being in 1948. One group's war of liberation and return to a homeland was another's loss of homes, villages, and farms. Grandparents stood with the children. A few carried keys to family homes that now house strangers. "I drive past my old family home as I go about my work. They have our house; I have nothing, not even citizenship." So says one man, a stateless resident of East Jerusalem. Most brought symbolic keys only, neck pendants or keychain ornaments.

Alongside these necklace or pocket keys people carried small metal depictions of Handala, a barefoot child whose head is spiky like a prickly pear cactus. This character, created by Palestinian cartoonist Naji Al-Ali, represents the cactus, *sabr*, for its persistence, deep-rootedness, and steadfastness in the face of difficulty. For Israelis the same cactus species is a symbol of the *sabra*, a Jew born in Israel, a person spiny on the outside, sweet on the inside. Cartoonist Kariel Gardosh gave the *sabra* its own cartoon character, the affable and gutsy Srulik. For both Israeli Jews and Palestinians the prickly pear is an important symbol of belonging to the land. Yet the plant itself is a colonist, native to Mexico and the southwestern United States. The cactus is common on the edges of fields or in the ruins of old Palestinian villages. It stands silent—every farmer I spoke with told me that it made no discernible sound—and so we project our own songs into the spines and pads and bring caricatures of its form to protests.

After the children left, the crowd grew to one hundred people clustered with their backs to the plaza's retaining wall. The security forces moved to block plaza exits. Banners were unfurled: "Returning," "All of Palestine is ours," "Let's go." A few Palestinian flags appeared from pockets and backpacks. A woman touched her cigarette lighter to a small polyester Israeli flag, scorching it in curls of black smoke. When three teenagers tried to carry the Palestinian flag through the gate, firm hands of Israeli security men repelled them. Then the sound: Handclaps, a rhythmic groove to move feet and voices. The chants: God is supreme! The right of return is sacred! The tempo of clapping surged; the crowd shouted louder.

Then, after thirty minutes of call-and-response, a policeman megaphoned the protesters, ordering them to disperse. Within a couple of minutes, the armed men slammed into the crowd. They knew whom they wanted: two men, headlocked and overpowered, were dragged to the armored truck on the street. Protesters moved to the stairways. A boy was hauled to the fence and arm-wrenched into a scream. A black-tunicked older woman went down, pushed as the security men surged, "punishing alike both those who were rioting and those who were not." She stood up, then limped through the Damascus Gate.

A few minutes later, a twenty-year-old Palestinian threw the cap of a water bottle at the security men. Their response was immediate. Thrown objects are intifada. Their ranks closed, then they broke, a spillage of well-aimed force into the throng. Palestinian medics pulled away the bruised and fallen. A scrum of Israeli security forces took down the lead medic and he passed out, head on stone. Several more surges from the perimeter fractured and dispersed the crowd. Half a dozen teenage girls regrouped and taunted the soldiers. The men responded with a chase and wrist grabs. The girls slipped away, continuing their pulsing chants. One soldier snapped, his face an agony of rage. As he was about to shoot a girl point-blank with rubber bullets, his comrades wrestled him back, pulling his hand from the trigger, restraining his anger with yells and body hugs. The girls sneered and shouted.

No one died or was badly hurt that day. On other days shocks of sound from stabbings and shootings pass through the olive branches. Every few months bodies pass as stretchers carry people to the street: Israelis stabbed as they pass through the gate, Palestinian assailants shot dead after their attacks. Because the Damascus Gate area is a nexus of conflict in Jerusalem, a plaza where many of the troubles of the Muslim Quarter of the Old City are manifest, the olive tree makes an occasional appearance in newspapers and television reports. It is not, though, a sylvan bystander. The fates of olives and people in this region are linked, joined by a reciprocal relationship that dates from before Judaism, Islam, and any of the modern nations in the region.

Within the hour after the Nakba Day protests, the market in front of the Damascus Gate resumed. Under it all, water dribbled over stone. The soldiers returned to West Jerusalem, a place of irrigated lawns, water parks, and fountains. The protesters, once they passed through the gated orifices of the Israeli separation wall, returned, old keys unused, to West Bank towns and refugee camps. The West Bank has been under military rule for fifty years, with some limited local governance after the Oslo Accords. Within this area fenced, guarded, Israeli-only settlement enclaves sit adjacent to Palestinian villages and farms. These settlements can be distinguished from the villages at a distance of many kilometers. Black water tanks crowd the roof-tops of the Palestinian villages. Waterworks on the Israeli settlers' roofs, often barely visible behind the irrigated palms, are spare and serve for solar water heating, not for survival. The twenty-first-century equivalents of Pilate's riot-inducing aqueducts run directly to these new developments, bypassing local towns and villages. Palestinian rooftop tanks are a buffer against shortage when the military and political conflict constricts water's flow, or for the yearly tightening of supplies as water flows abundantly to Israeli settlements but is rationed or restricted for others.

North of Jerusalem, on the Israeli side of the fence that cordons the West Bank, an olive plantation stands on the fields of Armageddon. The Bible's Book of Revelation fingers this land as the place where the armies of the end will gather: "Voices, and thunders, and light-nings. . . . The cities of the nations fell. . . . Every island fled away, and the mountains were not found." Olive oil with a foretaste of doom must sit well on some palates. Many of the trees in the plantation bore tags naming their Texan evangelical Christian sponsors, "Day-star Television Network: Planted for Mr. and Mrs. Tuck." Americans with an eye on the world's end are good for Israeli farmers. These foreigners picked up some of the costs of planting the orchard.

When I visited, late in 2014, during the November olive harvest, the only "voices and thunders" were the *gruu* of migrating cranes, the talk of farmers tending their machinery, and the sluicing tumult of millions of olives falling from harvester to hopper. Such agrarian sounds have dominated here for millennia. The Greek word *arma-geddon* derives from Hebrew *har Megiddo*, the hill of Megiddo. Seen from the olive groves, Megiddo is now a supine curve of sandy land. Its modern quiet belies a nine-thousand-year history as a city-state, a regional hub of agriculture, commerce, and government. Like Jerusalem, Megiddo rests on rock that has been tunneled, shafted, walled, and scooped. Megiddo derived its longevity and power from these waterworks, many of which comprise hand-chiseled conduits as tall as a human.

Now, rolled out at Megiddo's feet, thousands of meters of plastic pipe serve as the modern continuation of the stonemasons' work. A black tube threads every row of olive trees. The tubing lies on the dark, loamy soil, nestled close to the olive trunks, away from the wide, tire-rutted lanes that separate the tree rows. Every tube is perforated, a pencil-lead hole whose simple appearance belies its complexity. Leon Webster, the young man who runs this farm, stepped away from the olive-harvesting tractor to show me the workings of the irrigation pipes. He sliced open a piece of scrap tubing, revealing plastic boxes glued at regular intervals to the inner surface of the tube wall. Each box regulates the flow of water. When the field's control valves are open, water dribbles from each opening at a known rate. Drop by measured drop, the olives' shallow roots are bathed.

Israeli inventors developed flow-regulating plastic boxes and nipples in the 1950s. When combined with drip pipes from Australia and Europe, these new water emitters helped farmers in the young State of Israel make, in the words of Isaiah, restated in the country's declaration of independence, the parched lands and the deserts "rejoice, and blossom." Some of the irrigation water comes from reservoirs, some from diluted sewage effluent, and, in some places, from brackish seeps. In the Armageddon groves, sewage water from a kibbutz and prison flowed through the pipes. When rain comes only for a few winter months, what is seen elsewhere in the world as waste becomes a precious liquid. "Necessity is a good teacher" is an old

European aphorism that I heard repeated many times in my conversations with Israeli farmers and olive researchers.

Drip irrigation not only brought water to olive trees but also changed the nature of the trees themselves. The trees at Armageddon look nothing like the sinewy tree at the Damascus Gate or the furrow-barked ancients that grow on rain-dependent terraces. The olives here are young trees of new varieties, bred to grow fast and yield much oil. A harvesting machine as tall as a house—a modified grape picker—straddles the rows and roars as it thrashes olives from twigs to hopper. In the wet soil the shakedown uproots a few trunks, casualties that will be replaced by new saplings. No tree grows taller or wider than the harvester's mouth, about two meters tall and a meter across. Trees are planted an arm span apart, so their crowns merge, hedgelike. The Israeli man who pioneered this method, Lavi Shimon, was laughed at when he told international conferences that he grew olives in closely packed rows with irrigation lines, like grapes. Now, decades later, his innovative plantation methods have spread as far as Spain and Australia.

That olive trees respond so well to abundant water is a botanical curiosity, one that hints at the evolutionary history of the species. Most plants in arid lands receive extra water with gratitude expressed through vigorous growth and flowering. But the magnitude of their response is usually modest. Water vessels, leaves, and photosynthetic chemicals in all plant species are adapted to their homes, a fit that precludes other modes of existence. A wildflower that has evolved in shady woods will benefit from a little extra light, but it can never match the vitality of a cousin whose ancestry was in the full sun of the prairie. Watering a desert plant will slake its roots, but if transplanted to a moist bed, the plant's mastery of dry lands limits how much water it can process. Olives bend this rule. They are a species that grows in arid land but seems at home when transplanted to wetter soils.

Although olives now grow almost exclusively on dry sites in the modern Mediterranean region, this distribution is a fraction of their former range. Before they were domesticated about 6,500 years ago, olives lived throughout the lands surrounding the Mediterranean Sea for hundreds of thousands of years. Ice ages pulsed into the region, lingering for tens of thousands of years, then breaking into warmer interglacials before the ice returned. During the coldest times, olives persisted in unfrozen refugia scattered along the seacoasts. There they grew in the shelter of south-facing slopes and along waterways. In warmer periods the species expanded its range, carried by olive-eating wood pigeons. Part of this range included the arid hills, but many trees lived with willow and other wet-rooted species in riverine habitats. Like cottonwood or Callery pear trees whose ancestors' evolutionary experience prepared them for urban life, the olive species' past allows it to exploit new agricultural technologies. When drip pipes arrived in fields during the latter part of the twentieth century, olive roots were ready to thrive in the moistened soil.

The narrowing of the olive's range to dry lands was entirely caused by human activities. We reserve wetter areas for thirsty species such as citrus, grains, and vegetables. Their intolerance to drought gives them a place in the lushest fields. None of these other species can match the adaptability of the olive tree's roots and the drought tolerance of its leaves. So we have confined the olive to places where seasonal aridity is severe. There the olive is more abundant than any other tree species, a dominance brought about by the human love of oil.

Israeli farmers now produce an abundance of olives and olive oil. This productivity emerges from the union of the olive's physiological predispositions with drip irrigation technology, fuels for plastics and pumps, and a state strong and organized enough to capture and distribute water. Despite these many advantages, many Israeli olive farmers struggle to stay in business. Paradoxically, in a country founded on a religion in which olive oil is a central symbol, the biggest challenge for olive growers in Israel is cultural. Because Israel came into being largely through the recent immigration of a diaspora, olive oil can be a hard sell to kitchens whose familial diet is rooted outside the Levant. European Union agricultural subsidies compound the problem, stocking Israeli supermarket shelves with olive oil priced so low that local farmers cannot compete.

Israeli Jews consume, on average, about two kilograms of olive oil each year, one quarter the amount used by their Arab neighbors. When olive oil sales are weak, farmers leave the land, then developers and local governments convert abandoned olive groves to housing. Although Israeli law protects the olive tree, unmanaged orchards get overgrown and are vulnerable to fire. A vigorous blaze kills the olives, removing legal impediment. Matches are apparently not hard to come by. So Israeli researchers and government officials, in addition to fighting olive pests and developing new tree varieties, must bring oil back to the Israeli Jewish people. By combining a certification system to ensure quality with a marketing campaign to proclaim the health and gastronomic benefits of olive oil, Moses's "land of oil olive" might yet retain viable olive farms. Moses doesn't sell, though. Adi Naali, CEO of the Israeli Olive Oil Board, told me that although the promise of Armageddon can attract American money, appeals to Scripture are a poor way of getting olive oil into Israeli homes.

At the Damascus Gate a stub of iron pipe and a plastic drip tube protrude from the soil under the olive tree. In dozens of visits in spring, summer, and early winter, I never saw water dribble from the tubes' mouths. Instead, the tree feeds on rain and whatever vendors or tourists spill on the ground. Life inside the human throng at the Damascus Gate seems to suit the tree. Tresses of young stems, covered in deep green elongate leaves, hang from the stone gray fissured bark of the tree's older stems. Now, in November, dozens of black olives cling to every drooping stem.

Puut. Fruit drops onto the drizzle-wet pavestones, joining the foot-ground smear of earlier falls. Human hands have browsed all the reachable fruit, so the olives fall from the higher branches, their tight-skinned smack into the puddles sounding like the fall of the big, tree-swollen raindrops of the ceibo tree. I wrap a handful of olives inside a handkerchief and carry them back to the rooftop dormitory in which I'm staying. There, as a rooster tries to outsing the muezzins, I grind the olive fruits' flesh, staining my fingers purple as I force the fruit into a watery mash. A scum of pink oil veneers the surface, the remnants of the anthocyanin pigments in the ripe olives' skin. I taste the oil. It is rancid, with a sour, stale twist in the nose and a touch of bile on the tongue.

I dumped my pressing but learned a lesson about olive trees. Hundreds of olive pits roll underfoot at the Damascus Gate tree. In a few minutes' work I plucked a meal's worth of food, albeit an overripe one. Olive trees' yield can be prodigious, even when the water supply is meager and roots are confined between stone in shallow soil. Ever since early humans and our hominin predecessors ate wild olives, leaving pits for Israeli archaeologists to find tens and hundreds of thousands of years later, olive has been the species that turned the stony hills of the Levant into producers of energy-dense food. A bowl of olive oil contains twice as much energy as the same weight of meat. Olive oil production takes less toil and water than does livestock. Like hazelnuts in Mesolithic Scotland, olive fruits made the landscape hospitable to humans. In the Chalcolithic, the age of copper, farmers in the western Mediterranean discovered that productive trees could be propagated by slicing, then replanting trunk calluses or by snipping the whiplike sprouts that emerge from olive trunks and roots. Later, Greeks grafted desirable cultivars onto vigorous rootstock. Studies of the genetics of olive trees show that almost all the olive trees in the Mediterranean, whether in a "wild" area or in an orchard, descend from cultivars. Trees whose genealogy has not intersected with the hands of humans are very rare. The well-being and persistence of humans and olive trees have been conjoined for thousands of years.

Where water, irrigation pipes, and money are in short supply, farmers use the old ways of cultivating and harvesting olives. To glimpse these methods, I took the bus north from Jerusalem, traveling to olive groves and oil presses near Jenin, on the other side of the Israeli separation barrier from Armageddon. There I stayed with Palestinian farmers for whom rain was the only moisture available to their trees. Olive trees stood in rocky soil, each many meters from its neighbor. A few trees were a meter or more across at the base, planted perhaps one thousand years ago. Most were as thick trunked as a human chest, a few decades or a century old. The majority of the trees were Souri, a variety whose physiology is well adapted to long, rainless summers and shallow soil. Farmers called the oldest trees Rumi, a name for the Romans, whose hands may have planted some of these ancient trees. Farmers told me that they'd tried the new varieties of olive tree, but without irrigation these horticultural novelties withered. Where water is scarce, farmers do best by sticking to the olive varieties whose ancestors have succeeded for many generations on dry hillsides. The DNA of these trees presupposes no easy flow of water.

We picked by hand onto tarps spread below the tree. The olives left the field in the bed of a trailer dragged by a 1950s tractor. Donkeys stood in neighboring fields, carrying water to the workers and sacks of olives back to homes and presses. In the field the fall of olives from the hands of inexperienced visitors like me was a hesitating smatter. Farmers and their families were better pickers and raised unrelenting hailstorms of olive-tarp percussion as their fingers worked the twigs. Each tree was haloed by human voices. Through my rudimentary Arabic and the translations of my fluent companions, I heard fragments of conversation. Men on ladders argued about the best way to prune each tree, cook a sheep, and get the most from the local olive press. Women's talk fell mostly silent in the presence of male guests, but as soon as the foreigners moved to another tree, laughter and family news gusted through the branches.

During the harvest, every tree became a center of human storytelling, stories that comprise people, trees, land, and the relationships among them. By the time workers have picked a field, tens of thousands of words have flown from mouth to ear. Part of the landscape's mind—its memories, connections, rhythms—is thereby held in human consciousness. Work among the olive trees does more than yield oil; it creates and deepens the stories from which are made human and ecological communities. The olive tree near the Damascus Gate and the pear tree in Manhattan perform similar functions, providing shaded spaces where people exchange news and goods. In Manhattan more people are involved—thousands interact around the tree every day—but the city yields briefer contact than the intense knotting of conversation among families and close associates in a Palestinian olive grove.

Across the separation wall, olives are mostly picked by machines or by Thai "guest workers" brought to Israel to replace the Palestinians and agrarian Zionists who used to harvest, hoe, and dig. As in other industrialized nations, Israel's human social networks now get little help from trees. So few Israelis now work in agriculture that the 95 percent of Israel's kibbutzim depend on foreign workers. Almost no farmworkers speak Hebrew; agricultural knowledge about Eretz Yisrael dwells inside human minds in the Thai language. The Ministry of Agriculture now has a program to encourage Israelis to work on farms. There citizens learn from workers whose passports are foreign but whose knowledge of the land is native.

Unlike my journey to Armageddon, a car trip as uneventful as drives in Western Europe or America, to get to Jenin's fields I passed into a region where military control is present at almost every turn. My visits started before the 2014 Gaza conflict, continued through the time of most active bombing, and concluded months after the August cease-fire. At all times, crossing the "border" from Israel into the West Bank—a term not liked by Israeli politicians but used by their soldiers in the field—involves inching through diesel fumes for hours toward checkpoints, showing identity cards under the muzzles of automatic weapons, and driving on the broken road surfaces of a society where the upkeep of infrastructure is often financially and logistically impossible. Traffic jams were longer and soldiers more brusque when the war was most active, but even in quieter times the checkpoints evinced the unnegotiable authority of the Israeli military in the West Bank. On the other side of razor wire and walls, faster, Israeli-only roads with rapid-pass checkpoints run to the settlements. A journey that took two minutes in an Israeli vehicle took me two and a half hours the next day when traveling on routes designated for Palestinians. As I sat in traffic, waiting for soldiers to inspect the public bus, I saw olive trees spray-painted on the walls of the road running from Jerusalem to Ramallah. These were not the pastoral scenes of "biblical peoples" that adorn so much of Jerusalem's religious-tourist kitsch but grieving women holding uprooted trees. Deuteronomy's injunction to "not destroy the trees" during siege and war has not been well heeded during the last decades.

All of the Palestinian farmers with whom I spoke had lost trees to the Israeli army, the separation barrier, or settlers. Many had portions of their land sliced away by the barrier or by the fences of settlements. They showed me cell-phone photos of settlers cutting and herbiciding trees, shooting or beating people, and setting fire to orchards. Farmers complained of soldiers giving permits for only brief visits to the groves behind barriers. Some farming families received permits for the grandfather only, yet it takes three to four hundred hours of labor per hectare to manually tend and harvest an olive grove, so the stranded trees go to waste. At many gates, tools and water bottles are barred. Palestinian farmers must buy drinking water from Israeli settlers living on the farmers' former orchards.

It takes only a few years of this constriction to break the bond between trees and people. Groves get weedy and fire prone. Trees go unpruned. On average in the West Bank, the harvest for olive groves that are put behind barriers falls by 75 percent. Land that is not worked for three years can be claimed by the state. Land that is deemed a security asset will be taken by the army. If the Israeli government is displeased with the actions of Palestinian politicians, settlements that were previously classed as illegal are normalized. Piece by piece, the land of the West Bank is being annexed and its people driven into smaller reservations. Hope also gets less capacious. A farmer walking between rolls of razor wire, returning from his few remaining trees on the wrong side of the barrier, told me, "It could not be more miserable for us. . . . No matter what we say or do, it makes no difference." Formerly he traveled by tractor; now he trudges alongside an old donkey.

It does not take much work for demagogues to turn this frustration to political or military ends. At the Jenin refugee camp, a place of narrow streets and hastily built concrete homes, the only posters on the walls are those of men who died fighting the Israelis or who murdered others by killing themselves. At the Freedom Theater on the edge of the refugee camp, young men said that their only childhood wish had been to die while inflicting a wound. What Palestinians call the Israeli occupation, they said, was as much about occupation, colonization, and destruction of thoughts and dreams as it was about the land.

In Israel too dreams are cut back by violence. Hamas fires rockets at Israeli civilian targets from Gaza. Suicide bombs and knife attacks are seldom far from the headlines and, in a small country, seldom far from home. In a state born partly from European genocide and ringed by mostly hostile nations, the shadows of past and possible future annihilation are deep and real. Almost every Israeli I met during the Gaza war had a close family member serving in the conflict. Many car radios were tuned not to music or news analysis but to the frequency on which rocket alerts are broadcast. Under such continual assault and threat, fence building seems inevitable.

Israel and the lands that it holds under military control are home to the cultural equivalent of Amazonian ecology, endless, looping strands of conflict. The path to *súmac káusai* is hard to discern.

Near Jenin I met an American businessman buying olive oil. He was Jewish and had lost close relatives to Palestinian suicide bombers in Israel. His presence near Jenin, birthplace of a disproportionate number of bombers and fighters, seemed puzzling. But his answer to my queries was straightforward: he's not interested in the past; he wants to find and encourage the good for the future. He was in Jenin for a particular good, fair-trade olive oil, produced in partnership between Canaan Fair Trade and the Palestinian Fair Trade Association.

These organizations amplify what happens at an olive tree during the harvest. They reconnect atomized communities, restoring and reinventing some of the human social networks that underpinned all agriculture in the region. In the nineteenth century and before, Palestinian villages managed agricultural land through a process called *musha'a*. Each family's capacity to work the land determined the acreage that they tended. Assignments changed every year or two, as the human community changed. The Ottoman Land Code of 1858,

continued and modified by the British, then the Israelis, ended this
arrangement, imposing individual deeds, mostly to clarify taxation
structures. The Palestinian Fair Trade Association takes the old
spirit of communication and cooperation and translates it into the
modern economy. By pooling resources and acting in unison, farmers negotiate better prices, plan plantings, and work together to improve the quality of their oils. Canaan Fair Trade then links farmer
cooperatives to olive oil markets in America and Europe. Like the
Israeli marketers trying to build a local market, these West Bank
farmers and exporters know that pleasure on the palate can keep
trees and people on the land.

Here, then, might be an alternative to the Levant's geographies of
fear, the self-perpetuating cycles of death wishes and fences. There
are no posters of the dead on the walls at Canaan's olive-processing
warehouse near Jenin. Instead, the olive press sits inside walls tiled
with Arabic calligraphy:

جذور	Juthur	Roots
زيتون	Zaytoon	Olive
الذوق	Adh-dhawq	Taste
جمال	Jamal	Beauty
تعاون	Ta'awon	Cooperation
ماء	Maa'	Water

In what is now Syria, clay tablets with Ugaritic cuneiform script lay
buried for 3,500 years. Now unearthed, the god Baal can speak
through Canaanite scribes. He tells of "tree's words" and "whispers
of stone," the sounds of rain. In the autumn he is the cloud rider. The
tablets declare that when the land receives him as rain, war is removed from the Earth, love is set in the ground.

The Canaanite prayer "Look to the earth for Baal's rain" is an old
one, recorded not only on clay in ancient human libraries but also in

botanical cuneiform, traces of pollen left in the soil. These pollen records date back hundreds of thousands of years, revealing the climatic rhythms within which human cultures evolved, thrived, and, sometimes, failed.

Jerusalem sits on a limestone ridge between the Mediterranean coastal plain to the west and the Dead Sea and Jordan to the east. In springtime, when the tiny white flowers of olives open, hundreds on every tree, their yellow pollen catches the Mediterranean winds and sails east. From the tree at the Damascus Gate, pollen travels over the walls of the Old City, across the Kidron Valley to the Mount of Olives. It flies over the separation wall, villages, and settlements, then down from the hills and across the caves at Qumran. The pollen grains are now four hundred meters below sea level, at the Dead Sea. Here they either blow across the sea to Jordan or sink into saline waters. Every year more pollen and dust arrives, sprinkling the sea's bottom with a record of the springtime bloom. Over thousands of years, the sea's sediment and its trapped pollen built into layers meters thick.

The Dead Sea is now falling, drawn down by modest rainfall, excessive upstream withdrawals for irrigation, and large evaporation ponds for harvesting salt. As the water level drops, old seabed deposits are exposed. By drilling cores into these salty gullies, then paring away layers of old sediment, geologists and biologists reconstruct the past, just as they did with the Florissant paper shale. Similar cores from the Sea of Galilee corroborate the Dead Sea record for the most recent millennia.

Sediment and pollen deposits reveal the primacy of Baal. Changes in rainfall have caused the Dead Sea, and its Ice Age predecessor, Lake Lisan, to rise and fall by hundreds of meters in elevation over the last quarter of a million years. Some centuries have been soaked, others parched. Pollen records follow the rain: from lush to desert, then back again, and again.

These cycles are driven by forces on the other side of the world.

When icy meltwater pulsed into the Atlantic at the end of the last ice age, the colder ocean drove less heat and moisture into the Mediterranean. The rains stopped, the Dead Sea fell, and the land turned to desert. When the flow of ice water in the Atlantic was slow or stilled, the Levant rains again fell. In these moist times, people moved. The first hominin and humans to leave Africa traveled into and through the region, and adjacent Arabia, mostly during these wet intervals. The descendants of these first Levantine travelers and colonists—the first of the region's many waves of diaspora—are now the native peoples of Europe, Asia, Australia, and the Americas.

In more recent times, pollen records of domesticated crops, especially records of olive, show the waxing and waning of human culture in the region. In Dead Sea sediments from 6,500 years ago, olive pollen suddenly becomes more abundant, coincident with the species' domestication. In the early Bronze Age, about 4,000 years ago, the climate was humid and olives thrived. There followed nearly 2,000 years of modest oscillations in climate and vegetation until the late Bronze Age. Sediments from 1250 to 1100 BCE contain almost no pollen from olives or other Mediterranean trees. Then the sediments themselves disappear. The geologists' cores contain a blank space, the mark of an interruption to the orderly accumulation. The Dead Sea dropped so low that pollen landed on windblown dunes, not water. When the rains returned a century later, they fell on a land depopulated of humans and their domesticated trees. Archaeologists refer to the ensuing cultural upheaval as the "late Bronze Age collapse." Ugaritic texts of this time talk of grain shipments as matters "of life or death." Scribes from northeastern Africa and the Levant bemoaned famine. Ice from the north might again have been partly to blame. Before the drying of the eastern Mediterranean, ice coverage over Greenland was at a peak. The ice sheet then partly melted, bringing drought on the other side of the world.

That Greenland is now once again melting will likely not help the agricultural future of the Middle East. When the World Resources Institute ranked countries by projected levels of "water stress" in the coming decades, Israel and Palestine were in the highest tier in all three sectors: agriculture, industry, and domestic use. The Dead Sea is close to the point at which pollen will again fall on dust, not water.

Elijah is purported to have defeated Baal, the false god of the Canaanites. But Baal's hand is evident in the orchard piping of Armageddon and in the water-thrifty genetics of old West Bank trees. He reveals his power in irrigated settlements and droughted Palestinian villages. When the army bars agricultural workers from carrying water across the separation barrier, it is borrowing Baal's strength. In the marketplaces too we hear his name. In both Palestinian and Israeli markets, fruits and vegetables grown without irrigation are called Baal but, in my limited experience and in the more extensive research of linguist and historian Basem Ra'ad, no vendors will acknowledge a connection to the pre-Abrahamic god.

Look to the earth for Baal's rain. An ecologically apt prayer, and perhaps one of desperation. The correlations among records of rainfall, pollen, and the fate of human societies seem to argue for fatalism: "No matter what we say or do, it makes no difference." Yet the pollen records show that Baal's moods, while powerful, do not entirely determine the fate of people and trees. Olive pollen persisted through a dry period around three thousand years ago, evincing the work of a poorly known culture of successful arid-land orchardists. Likewise, the desires and skills of the Greeks, Romans, and Byzantines turned the thin eastward drift of pollen from the Judean highlands into fat clouds, thickened by the pollen of grapes. A taste for oil and wine, combined with a knack for irrigation and centralized planning, turned the hills into orchards and vineyards. The climate helped—lake levels were often high in Roman times, indicating good rains—but even in dry periods abundant olive pollen streamed from the hills. Pilate's Jerusalem aqueducts, one of many Roman water management projects, renegotiated the terms of people's relationship with Baal. Conversely, other periods had a good climate but little olive pollen. Pollen levels dropped sporadically during the more lush parts of the Bronze Age, when war and political instability prevented people from working with their trees. Then, during the late Iron Age, about 750–550 BCE, olive agriculture in the thriving kingdoms of Judah and Israel was undone by Assyrian and Babylonian invasions. Look to Baal, yes, but if the social context is riven, farmers and trees cannot twine and thereby yield food from the land.

Like animal communities in Amazonian bromeliads, the roots of boreal fir trees, and Callery pears on Manhattan's streets, the olive groves of the Levant depend for their vitality and persistence on stable relationships with other species. In the olive's case, the most important species in the tree's network is *Homo sapiens*. Severing these relationships kills just as surely as cutting individual trees. Bronze Age wars, Babylonian invasions, and modern separation barriers override Baal's generosity, withering a potentially fertile land when people and trees lose the connections that give life to both.

War and dislocation of human communities do more than cut relationships in the present moment. Exodus of people from the land erases the embodied knowledge of a place. Amazonian Waorani displaced by industry, North American Indians killed and evicted by colonists, Judahites removed to Babylonian exile, Palestinians after the *nakba*, and even the peacetime depopulations caused by the unprofitability of farming: all these cause the memories embedded in the connections between people and other species to fall out of existence. Exiles can write down and preserve what we carry in our minds, but knowledge created and sustained by ongoing relationship dies when connections are broken. What remains is a network of life that is less intelligent, productive, resilient, and creative.

We inherit and live within these dislocations and losses. In remaking connections, though, we stitch life back together, increasing the network's beauty and potential. In Ecuador the Omaere Foundation takes degraded land and cultivates both plants and people's relationships within the botanical community, taking strands of knowledge from grandparents and passing them to hundreds of young people. Omaere's cofounder, Teresa Shiki, a Shuar woman, told me, "Put away that notebook. What is written dies; only what you live in relationship survives." When the New York City Department of Parks and Recreation involves neighborhood residents in planting trees, people gain living connections that, although they are much less diverse than those in Amazonian forests, yield better life for trees and people. In the boreal forest, conversation among former adversaries in political struggles draws together strands of lived experience, a network of thought emerging from forest life. Canaan Fair Trade, the Palestinian Fair Trade Association, and the Israeli Ministry of Agriculture each try to encourage networks of relationship and conversation through which the conjoined lives of trees and people can be understood, remembered, and tended.

Amid the sound of sirens, a Hamas rocket flies over Gaza's walls toward Jerusalem. The rocket is one of the thousands of explosives slung by both sides of the conflict in July of 2014. Around the olive tree at the Damascus Gate, the sirens do not slow the thronging Ramadan market. People stand shoulder to shoulder in the crowd and jostle as they push toward the vendors who have squeezed into every part of the plaza. The olive tree is trussed with ropes holding a canopy of tarpaulins over the crowded market stalls and walkways. The covering gives shade during the day but also serves as a shelter from dust and wind at night. The war has slowed people's movements through separation-barrier checkpoints, the Palestinian market vendors tell me, so anyone with a home on the other side of the barrier prefers to sleep here, under the tree.

To display their fruits, vendors have made a table from stacked cardboard boxes. These containers are marked with names of Israeli agricultural companies whose orchards are located along the coast, away from the disputed lands of Jerusalem and the West Bank. Like the olives of Armageddon, the plums and oranges in this market came from fields made to "rejoice, and blossom" with drip pipes. Here in East Jerusalem, though, there is no water for irrigation. Israeli water-supply policies have cut the flow of water to tens of thousands of people. Some residents must buy their drinking water in bottles. With the exception of a few olives from local rain-fed trees, every fruit in the market is a small packet of imported water.

Before sundown, men from a Qatari charity stand at the open doors of a van and hand food to the crowd, a *sadaqah* offering to the needy. Like the fruits in the market, each boxed meal carries with it water from elsewhere, a Pilate's aqueduct sustaining life in Jerusalem. The sun sets and within minutes the crowd is gone, headed inside to fast-breaking *iftar* feasts. A few vendors remain in the near-empty plaza, opening their boxed meals. As dusk advances, a black tomcat, a ginger, and a tabby slide out of walls and gather silently around the fissured trunk of the olive tree. There, as their ancestors have likely done for millennia, the cats feed at small piles of scraps left by market vendors, the fickle leavings of Baal's largesse.

Japanese White Pine

Miyajima Island, Japan
34°16'44.1" N, 132°19'10.0" E
Washington, DC
38°54'44.7" N, 76°58'08.8" W

Juniper and pine logs spit and smolder under an iron cauldron. Years of smoke from the fire have blackened every surface in the room. Stalactites of oozy carbon hang from the ceiling and smear the roof-anchored cable that holds the vessel over the fire. Wooden walls and benches smell of resinous charcoal and ash. To enter the room, I duck through a low doorway, under the carved wooden sign for Reika-do Eternal Fire Hall. As I step over the threshold, I see clear air flowing around my feet, then sweeping up to the fire at the center of the room. Eye-burning smoke curls around the cauldron and twists back to exit the unvented room through the upper half of the door opening. The fire's exhalation disgorges from the door, up around the hall's curved pagoda eaves and into mountain air.

Inside, the smoke haze is so dense that coughs and voices seem muffled in smog. Pilgrims and tourists, lungs hacking, peer at the fire, holding tea bowls. We heft the lid from the cauldron's lip, loosing a boom from the barrel-size metal pot as we move the slablike covering, and dip our bowls into the warm water. A few sips are reputed to heal all physical ills, but even without its purported supernatural properties, the water tastes sweet after our five-hundred-meter climb from the sea to the mountain peak on which the hall stands.

Wood has burned here for 1,200 years, a log-stoked fire kept aflame by the followers of Kōbō-Daishi, founder of Shingon Buddhism in Japan. After his studies in China in 806, he spent one hundred days in ascetic practice on this mountaintop on an island in the Seto Sea, just off the coast from Hiroshima. His campfire—Kiezu-no-hi, the eternal flame—has burned ever since, thanks to the monastic community that he founded. Water heated by the fire not only heals but also is purified by the flame. During the Edo period, the seventeenth through nineteenth centuries in Japan, monks carried water from the fire down the mountain to the temples below. There they used the water to make inks for copying the sutra, Buddhist sacred texts. Kōbō-Daishi was one of the many religious and political leaders who chose the island for their sacred buildings. The Reika-do Eternal Fire Hall sits among dozens of Shinto and Buddhist shrines and temples. The island gets its formal name from the main temple, Itsukushima, but is often called Shrine Island or Miyajima.

The shrines brought me here, but not as a typical pilgrim. My aim was botanical, to visit the original home of a tree that now resides in Washington, DC. The tree's journey owes its first steps to Miyajima's shrines. By sanctifying the forests that surround them, sacred buildings give plants from the island powerful cultural valence in Japan. This is especially true in the Shinto religion, where boundaries among humanity, the spirit world, and "nature" are illusions that special places like Miyajima can help us transcend. Forests around Shinto shrines, comprised of sacred *shinboku* trees, are the nexus of human and nonhuman, the living and the dead, the spirits and the physical world. On Miyajima the whole island is sacred, a living shrine to the connectivity of the world. Just as the roots of all trees integrate the ecological community, trees in a *shinboku* forest integrate the many dimensions of the Shinto universe, ecology included. The tree whose place of origin I seek is a Japanese white pine. In 1625 the sapling was dug up and taken to the mainland. There it was grafted onto the roots of the more hardy black pine and gradually sculpted into a bonsai tree. Untended by humans, the tree could have grown to twenty meters, as large as the ponderosa that shaded me in Colorado. But regular pruning guided the tree to a size that would, were I to stand next to the plant's ceramic container, cast shadow on my knees and no higher. Pruning of branches and roots not only dwarfed the tree but also shaped it into an upright form with a balanced dome of needles. As with many bonsai trees, wires wrapped around branches further nudged the tree into a shape pleasing to the human eye.

The ponderosa pine's roots and their fungal network could draw water from deep in the soil, far beyond the tree's outer branch tips. For the grafted Japanese white pine, as with all bonsai trees, the absence of this large catchment necessitates close attention by humans who daily, sometimes twice daily, drizzle offerings of water. The caretakers also, through yearly or biennial removal of older roots, maintain a haze of young rootlets within the confined space of the wide, shallow planter. Although mutualistic fungi do colonize the soil in bonsai containers, human labor mostly replaces fungal work.

For 350 years one family, the Yamakis, cared for the tree, continuing for many generations the work of their ancestors. In 1945 the garden wall of the Yamaki family compound in Hiroshima saved the tree from the blast of the atomic bomb. The home was three kilometers from the epicenter, so walls stood, even though windows exploded and lacerated members of the family, who were all inside. The tree remained in Hiroshima until 1976, when it made the reverse journey of the *Enola Gay*. The Yamaki family and the Japanese government presented the tree to the United States to mark the country's bicentennial.

The Yamaki pine now lives in the National Bonsai & Penjing Museum at the U.S. National Arboretum, in a suburb northeast of the U.S. Capitol. For a bonsai, the tree is large. Its ceramic planter is an arm length wide and a hand span deep. Twist-creviced bark wraps a trunk as tall as my forearm and as wide as a slender torso. The bark's calluses and accretions speak of the tree's longevity, drawing my gaze into flaked layers and clefts. A domed, flat-bottomed crown of needles sits symmetrically atop the bole. The arc of the top surface of this crown emerges from the billow of several branches. Their slope is just a little too lively to evoke a pastoral hill, yet the curve remains tender. My eye quiets on the sculpted living form.

Around the tree sit other ancients, trees from the eighteenth and nineteenth centuries. None is as old as the Yamaki pine, though, or from a more august germination seedbed. According to the records kept by the Yamaki family and now the National Arboretum, the Japanese white pine in Washington started life in the early 1600s on Mount Misen's slopes, near Kōbō-Daishi's fire on Miyajima.

I leave the smoke of Reika-do and continue my search for the place of origin of the bonsaied white pine. The forest's plants evoke in me a feeling that reality has slipped. All is familiar. Wind in Japanese oaks and maples sounds as it does in the Americas: coarse grained and deep voiced in oaks, sandy and light in the thin-leaved maples. Yet when I attend to a visual detail—the contour of a leaf, the runnels in bark, the hue of a fruit—I'm unmoored by strangeness. This is no aftereffect of shrine smoke. My mind is foundering instead in the geographic manifestation of plant evolution's deep history. The plants of East Asia, seemingly so far from eastern North America, are in fact close kin to the plants on the mountain slopes of Appalachia, closer kin by far than the plants of the northwestern United States, or of Florida, or the arid lands of the Southwest. On Miyajima I walk with sumac, maple, ash, juniper, pines, fir, oak, persimmon, holly, beautyberry, blueberry, and rhododendron. A few Asian specialties spice this thoroughly Appalachian community, curiosities like Japanese cedar, snake vines, and umbrella pines. The cedars intermingle their soft, extended sighs with the more familiar sounds of oak and maple. These few uniquely Asian species aside, almost every plant I encounter has the countenance of an intimate. When I approach, though, the details of the visage confound me: needles seem oddly splayed, acorn cups are too delicate, and berries cluster in strange geometries. The taxonomy of these plants, inferred from DNA, fossils, and plant anatomy, shows that the Japanese species are siblings of plants in the Appalachian forests. My senses are struggling at a family reunion composed of the sisters and brothers of people whom I know well.

Modern geography gives little cause to expect that plants from eastern Asia and eastern North America should have such tight familial ties. But the mild, humid climate in which temperate forests grow was once much more widespread, especially during and after the period in which the forests of Florissant grew. The Northern Hemisphere was swathed by forests composed of the ancestors of plants in modern East Asia and Appalachia. A colder climate and increasing aridity in the center of North America later fragmented this forest. Ice ages added deeper cuts, pushing temperate species into yet smaller southerly refugia. Climate change broke up and dispersed the family.

My trek to the centuries-old birthplace of the Yamaki pine therefore took me further back in time than the pine tree's germination. Here was the living legacy of what I had experienced through dead fossils at Florissant, a botanical signature of forest kinship reaching back at least thirty million years.

What I did not find among the Miyajima plants, though, were any Japanese white pine trees. Red and black pines crowded the mountain summit, but not one trunk or sapling had the characteristic five-needled leaf clusters of the species I sought. Japanese botanists and Western visiting scientists confirmed what I found. As far as we know, the only Japanese white pines now on the island live in bonsai pots or are planted in public spaces. Either the Yamaki oral history is embellished, perhaps because early plant collectors wanted the ca-chet of Miyajima for their bonsaied trees, or the island is not as it was four hundred years ago when the tree was dug, grafted, then potted.

Creative nomenclature cannot be discounted. No rules of bonsai horticulture forbid naming cultivars for noted landmarks, even when provenance is from elsewhere. Today tourist brochures and posters make liberal use of Miyajima's name. Representations of the island's plants, gates, and shrines are common. In just one block of shops in Hiroshima I saw miniaturized reproductions of shrines, island maple leaf pastries, shrine photographs enameled onto trinkets, illuminated *torii* gate decor on oyster-vending carts, and elegant woodblock paintings of temples. No doubt such appropriations also happened in the past. Miyajima may be therefore only a moniker for the tree, not a record of its origin.

Yet it is also possible that the Yamaki family's oral history holds reliable memories. Any family diligent and attentive enough to give daily care to a tree for centuries is perhaps likely to devote the same care to the tree's story. What was passed from mouth to ear might instruct us about the changing geographic range of the Japanese white pine over the last four centuries. Scientific records corroborate the story. The Little Ice Age spans the early years of the seventeenth century, the time when the Yamaki pine seed would have fallen from its cone. Written chronicles of the dates of river freezes and the timing of cherry blossoms, combined with information from tree rings and pollen, reveal an icy Japanese climate. Archives from Europe confirm the global reach of the Little Ice Age. In East Asia the centuries before this time, years in which the Yamaki tree's ancestors would have grown to maturity, were also colder than the present day. Pine pollen was then more common. Oaks and other broad-leaved trees were in decline. So although Miyajima may now be a little too warm and southerly to host the Japanese white pine, the island lived in a different climate four hundred years ago. When bonsai gatherers climbed the island's slopes in the 1600s, they would have found a forest like that of modern Japan's higher, colder mountains. According to the tree's *kleos*, sung over hundreds of years, these climbers found white pines on Mount Misen.

Trees in Miyajima grow surrounded by the sound of prayers. In front of shrines, wooden balls the size of oranges are threaded onto loops of rope that pass over pulleys mounted to the shrines' doorways. When pilgrims haul the ropes, the balls rise with the rope, then overtop the pulley and fall against one another, clacking and delivering a sonic petition to the shrine deities. Hands shake boxes of divination sticks, predictions of the future emerging from the shimmer of thin sticks colliding, a seethe of friction and clatter. Paneled, carved temple walls echo the double claps of hand palms beating in supplication and thanks. Pendent logs, hung from stout ropes, swing into metal bells, the logs' slam-splayed wood fibers velveting their strike. Coins fall into tympanic wooden offering boxes. All these sounds call forth the *kami*, the spirits, who dwell in the forest and its shrines. Human agency in these sanctified spaces acts largely through sonification of wood; vibrating cellulose is the mediator between our invocations and the spirit world. Spirits here live not in a removed heaven but within trees, the forest, and wooden shrines. Percussive wooden sounds draw *kami* from their homes in Earth's heartwood. From forests such as these the Yamaki pine moved to Hiroshima. There it outlived the age of lanes trundled by handcarts and horses, into years of fossiliferous cough and bombus of engine, to the froth of fast tires on asphalt. In 1945 the "big sound," the *BOONG* recalled by survivors, shook the pine.

Now, in Washington, helicopters fret the sky and the noise from distant expressways edges its way into every corner of the pavilions in which the bonsai collection is housed. Miyajima's forests are present in the bonsai pavilion too, in the cottony zephyrs of sound issuing from plantings of unpruned, full-size Japanese cedars and maples. The demeanor of Miyajima pilgrims also finds a place here, in the movement and voices of some museum visitors. They pause before the tree, bow to read the specimen label, then stand erect, receiving what the tree has to offer to their senses. Murmured commentary sometimes follows, musings about origins of the trees, visual balance in foliage, or underlying form in trunks and branches. After a minute the admirers turn and continue their walking meditation. Most visitors, though, bring acoustic vigor unknown among shrine pilgrims. As if at a fairground, the tree viewers bounce and sway, flinging their attention and laughter with haphazard enthusiasm. No tree receives more than a few seconds of their gaze, a process that concentrates aesthetic appreciation into an elemental state, a rapid sensory engagement with the tree. Astonishment, glee, and puzzlement pop from these encounters: shouts of disbelief at the age of the trees, exhortations for companions to marvel at shape or color, and questions thrown to the air about how these curiosities came to be.

For museum visitors bonsai trees seem to open a conversation in ways that other trees usually do not. Context is perhaps part of the cause: curators present the trees in a way that invites viewing. But in addition to the social signals given by containers, labels, and institutional sanction, the form of bonsai encourages connection. By bringing entire trees to the scale of a human head or torso, it allows our senses to, often for the first time in our lives, encompass and apprehend a whole tree. Like the pink-trousered girl of Florissant, visitors are starting or continuing a process that opens them to the lives of other species. When children gather at the Yamaki pine, crying out as they see hundreds of years of plant growth in a body as small as their own, a connection is made that lodges firmly in body and mind.

Air brings all these hundreds of years of sound from temples, forests, and cities to the needles, roots, and trunk of the Yamaki pine. The tree inhales and stills the air's fibrillating breath, holding it in wood, like a *kami*. Each year's growth ring jackets the previous, capturing in layered derma precise molecular signatures of the atmosphere, timbered memories. Wood emerges from relationship with air, catalyzed by the flash of electrons through membranes. Atmosphere and plant make each other: plant as a temporary crystallization of carbon, air as a product of 400 million years of forest breath. Neither tree nor air has a narrative, a telos of its own, for neither is its own.

For the air, the tree, and the forest, form and narrative emerge from relationship. Selves are ephemeral aggregations, made of the enduring substances of life—connections and conversations. Into these relationships steps the human, with shovel, clippers, and a ceramic bonsai planter. The practice of bonsai seems at first to be a concrete manifestation and a metaphor of humanity's escape from the networks of life. Using manufactured blades, we appear to impose our own teleologies on all others. Through root pruning, trimming of branches, grafting, scoring of bark, and reworking of soil, the bonsai tree is captive to a master from whose mind the tree's future will emerge. Such might be one conclusion to draw from the Yamaki pine, the atomic tree. A plant turned into chattel thrall, then bombed.

The reactions of visitors to the Yamaki pine undermine such an interpretation. Bonsai does not escape life's network. Instead, like olive groves, bonsai trees bring to the surface what is harder to discern elsewhere: that human lives and tree lives are made, always, from relationship. For many trees it is nonhuman species—bacteria, fungi, insects, birds—that are the primary constituents of the network. Olive and bonsai trees bring humans to the center, giving us direct experience of the importance of sustained connection.

Should these connections break, life is diminished, sometimes ended. In the Levant, severance of relationship causes the decline or death of oil-bearing trees, along with the economies and cultures that depend on them. In bonsai, trees isolated from contact with people quickly die, carrying with them the fruits of centuries of tree growth and human work. These losses have fewer consequences for human pantries and family budgets than those in olive groves, but they strike deep within culture nonetheless.

Chinese and Japanese horticulturists have been aware of the primacy of relationship for centuries. *Sakuteiki*, the eleventh-century Japanese manual of gardening, possibly the oldest written record of landscape design, exhorts people to open themselves to the disposition of mountain streams, to wind and emotion. The author, probably Tachibana no Toshitsuna, son of an imperial regent, urged gardeners to bring into consciousness "wild nature," by which he meant not a separate nonhuman world but the inner nature of humans, other species, water, and rocks. This inner nature is animist: rocks have desires, trees are imbued with Buddha-like solemnity, and relationships among seemingly separate elements of the landscape—the arrangement of stones and plants, for example—govern the mood of the spirits that dwell among all things. He believed that both direct experience of the world and contemplation of the work of "past master gardeners" were necessary, a humble opening of the self to the knowledge of others. The garden is not an escape into domineering control of nature; rather it requires sustained attention to the networks of life, including the understanding of these networks carried in human memory. In the *Sakuteiki* careful listening results in an aesthetic attuned to the many relationships present within the garden.

Attention to the inner "nature" of the nonhuman, combined with close listening to the accumulated knowledge of many human generations, continues in later Japanese horticultural treatises. Shingen's fifteenth-century writings and sketches on landscape design, sometimes attributed to the eleventh-century priest Zōen, insist that "if you have not received the oral transmissions [teachings from older masters], you must not make gardens." A gardener must integrate this knowledge with "full attention" within the garden to the orientation of rocks, the movements of birds, and the form of tree limbs. Reverence, respect, and attention are the author's themes, not control.

Contemporary bonsai practice emerges from these philosophies. Standing next to the Yamaki pine that is now in his care, Jack Sustic, curator at the National Arboretum, spoke of listening to mentors and of giving attention to the trees, the same themes that Tachibana no Toshitsuna, Zōen, and Shingen wrote of more than half a millennium ago. Years of work with bonsai trees will, Sustic said, decenter a person, drawing the locus of attention away from the self. "It's less about me, much more about the tree and the work of the people who came before. This affects the rest of my life. I'm more tolerant, understanding." Sustic's career in bonsai began with an experience that Zōen might have called "full attention" and Iris Murdoch an "unselfing" into aesthetic experience. While serving in the army in Korea, he saw a bonsai collection from a bus window and lost for a moment his sense of time and place. "That's what good art does," Sustic said.

Aarin Packard, then the collection's assistant curator, talked to me as his fingers fluttered and groomed the bonsai trees' soil and twigs. Beginners believe that they can see far into the trees' futures, he told me, that they can impose form on trunks and branches. But as you learn more, you understand that form emerges from an unpredictable meeting of lives. "The great living American practitioners can see perhaps fifteen years into the evolving form. John Naka and a few other masters could see half a century, no more."

The future, the unfolding telos, is not contained in any self, in a tree seed or human mind, but has its origin and substance in living strands of relationship. Through horticulture, bonsai mirrors the nature of trees. A tree is the common life, a being that is multiplicity of conversation.

The indivisible atom turned out to be an illusion, one that ended six hundred meters above Hiroshima. The mask of individuality shattered, revealing unspeakable energy. At the temples the stone faces of Buddha melted.

In 1964 bomb survivors took fire from Kōbō-Daishi's pine logs and kindled a gas flame amid the cenotaphs and mass graves of the city's Peace Park. Wooden strikers on metal bells echoed the sounds of Miyajima's forest shrines.

From beyond the atom, the Yamaki family gave their art, a merger of lives.

Acknowledgments

My gratitude, first, to Katie Lehman, for inspiration, counsel, and encouragement as I worked on the manuscript, and for walking beside me among the trees, sharing their beauty.

I've had the honor and pleasure of working on this book with extraordinary colleagues. Not only did my editor at Viking, Paul Slovak, help me to imagine and shape this project, but his incisive and discerning readings and guidance improved the form and substance of the book. I also benefited greatly from Alice Martell's perceptive analysis of the book's ideas and structure. In her outstanding work as my agent, she guided my ideas to maturity, gave me unflagging support, and brought the book into being. Conversations with Kevin Doughten formed me as a writer during our work together on *The Forest Unseen* and, in the first stages of my explorations of future works, clarified my thinking about biological networks. I thank Viking's editorial, design, production, and marketing staffs for their fabulous work, especially Haley Swanson for editorial help and shepherding the manuscript, Hilary Roberts for perceptive and careful copyediting, and Andrea Schulz, Tricia Conley, Fabiana Van Arsdell, Kate Griggs, and Cassandra Garruzzo for editing, production, and design.

I thank Viking, the John Simon Guggenheim Memorial Foundation, the American Museum of Natural History, the St. Catherines Island Research Program, the Edward J. Nobel Foundation, and the University of the South for financial support. John Gatta, University of the South; Thomas Levenson, Massachusetts Institute of Technology; Barbara King, College of William and Mary; and Mike Webster, Cornell University provided generous support and advice in the early stages of the project. Rivendell Writers' Colony was a productive place in which to write and I thank Carmen Toussaint Thompson for all her help. I thank Sarah Vance for her advice, support, and practical assistance in the early years of this project.

I offer many thanks to all those who shared their insights and advice with me about particular parts of my work, and to all who extended hospitality to me during my travels: Buck Butler, Jon Evans, Mark Hopwood, Katie Lehman, Leigh Lentile, Deborah McGrath, Stephen Miller, Sara Nimis, Tam Parker, Greg Pond, Bran Potter, Cari Reynolds, Gerald Smith, Ken Smith, and Christopher Van de Ven, University of the South; Paul Becker, Carl Becker & Son, Chicago; Rex Cocroft, University of Missouri; Dan Johnson, Duke University; Pedro Barbosa, University of Maryland; Adrienne Christy, Metropolitan State University, Denver; Peter Matthews; Jonathan Meiburg; Paul Miller; Greg Budney, Cornell University; Randa Kayyali, George Washington University; Todd Crabtree, Tennessee Department of Environment and Conservation; Bill Kupinse and Peter Wimberger, University of Puget Sound; Martha Stevenson, World Wildlife Fund; Lang Elliott, Music of Nature; Dustin Williams, Williams Fine Violins; Joseph Bordley; Mariane Tyndall; Sanford McGee; Anna Harding; Laurie Perry Vaughen; Paddy Woodworth; Matt Farr, the Nature Conservancy; Richard Hofstetter, Northern Arizona University; Deborah G. McCullough, Michigan State University; Jeff Brenzel, Derek Briggs, David Budries, Susan Butts, Peter Crane, Michael Donoghue, Ashley DuVal, Justin Eichenlaub, Jon Grimm, Chris Hebdon, Shusheng Hu, Valerie Moye, Rick Prum, Sayd Randle, Scott Strobel, and Mary Evelyn Tucker, Yale University. Classroom conversations with students in my biology and literature classes at the University of the South enriched my thinking and writing. Jim Peters and Tom Ward generously shared their friendship and conversation, deepening and expanding my thinking through their example and counsel.

Ecuador: Esteban Suárez, Andrés Reyes, Consuelo de Romo, Diego Quiroga, Pablo Negret, José Matanilla, María José Rendón, Mayer Rodríguez, Ramiro San Miguel, and Kelly Swing, Universidad San Francisco de Quito and the Tiputini Biodiversity Station; the Ecuador Ministry of Environment; Eduardo Ortiz, René Bueno, Gladys Argoti, Lee L'Hote, Melissa Torres, Lauren Ostrowski, and John Lucas at the Institute for the International Education of Students and all the student and faculty participants in the IES Quito/Tiputini seminars, especially Given Harper for his friendship and ecological insights. Chris Hebdon, Yale University, assisted in many practical ways, sharing his extensive knowledge through both conversation and draft manuscripts, thereby helping me to understand the significance of what we heard in Ecuador. I owe particular thanks to Chris for clarifying and extending my understanding of the many ways in which modernity manifests across cultures. He also elucidated for me the many political and pragmatic dimensions of the choices people make as they express their culture to outsiders. Members of indigenous communities in and around the Amazon were extraordinarily welcoming and generous. Regrettably, these people and communities are sometimes subject to political persecution, so I offer my thanks but will not list names.

Ontario: Jeff Wells, Boreal Songbird Initiative; Phil Fralick, Department of Geology, Lakehead University.

St. Catherines Island: Royce Hayes, Christa Hayes, Jenifer Hilburn, Tim Keith-Lucas, Lisa Keith-Lucas, Jon Evans, Kirk Zigler, Ken Smith, Bran Potter, Gale Bishop, Mike Halderson, Eileen Schaefer, Arden Jones, the students in the University of the South's Island Ecology Program, and the staff and interns of the St. Catherines Island Sea Turtle Conservation Program.

Scotland: Laura Bailey, Edward Bailey, and Julie Franklin, Headland Archaeology; Rod McCullagh, Historic Scotland; John Gardner, Forth Energy; Donald Dalton; Jean and George Haskell; Jim Cornwall, National Mining Museum.

Florissant, Colorado: Jeff Wolin, Hebert Meyer, and Aly Baumgartner, Florissant Fossil Beds National Monument; Toby Wells.

Denver, Colorado: Laurina Lyle, Project WET; Rick Sargent, Sargent Studios; Matt Bond, Denver Water; Jolon Clark, Greenway Foundation; Casey Davenhill, Cherry Creek Stewardship Partners; Cynthia Karvaski, William "Pat" Kennedy, Jon Novick, and Ted Roy, City and County of Denver; Devan McGranahan, North Dakota State University.

New York: Hailey Robison, Warner Watkins, Stanley Bethea, Ofelia Del Principe.

Israel and the West Bank: Zohar Kerem, Jeff Camhi, the Hebrew University of Jerusalem; the Sisters and volunteers at Ecce Homo Convent, Jerusalem; Fred Schlomka, Mohammad Barakat, Yamen Elabed, Bruce Brill, and Yahav Zohar, Green Olive Tours; Adi Naali, Ibrahim Jubran, and Rowhia Ganem, Israel Olive Board; Leon Webster; Neta Keren; Ayala Noy Meir; Monaem Jahshan; Mohammed Al Ruzzi, Haj Bashir, and Majed Maree, Palestine Fair Trade Association; Nasser Abufarha, Manal Abdullah, and Mohannad Ghannam, Canaan Fair Trade; Maxine Levite, Michal Productions; Adam Eidinge. West Bank farmers and others asked not to have their names printed, a request that was perhaps the shadow of the unfulfilled requests for names that I received from Israeli security forces. I offer my thanks for my hosts' hospitality, shared work among the trees, and fruitful conversations.

Washington, DC, and Miyajima, Japan: Jack Sustic, Aarin Packard, and Avery Anapol, National Bonsai & Penjing Museum, U.S. National Arboretum; Felix Laughlin, U.S. National Bonsai Foundation; Brent Hine, UBC Botanical Garden and Centre for

Plant Research; Iwao Uehara, Tokyo University of Agriculture; Tom Christian, Royal Botanic Garden Edinburgh; Hiromi Tsubota, Hiroshima University; Farrand Bloch, Bonsai Focus; Peter Chan, Herons Bonsai; Derek Spicer, Kilworth Conifers; Rebecca Bates and Rob Foster, Berea College; Jordan Casey; Miki Naoko; Bruce Taylor.

Bibliography

Preface, Mitsumata, Maple

Basbanes, N. A. *On Paper: The Everything of Its Two-Thousand-Year History.* New York: Knopf, 2013.

Bierman, C. J. *Handbook of Pulping and Papermaking,* 2nd ed. San Diego: Academic Press, 1996.

Ek, M., G. Gellerstedt, and G. Henricksson, eds. *Pulp and Paper Chemistry and Technology.* Vols. 1–4. Berlin: de Gruyter, 2009.

Food and Agriculture Organization of the United Nations. "Forest Products Statistics." 2015. www.fao.org/forestry/statistics/80938/en/.

Goldstein, R. N. *Plato at the Googleplex: Why Philosophy Won't Go Away.* New York: Pantheon, 2014.

Knight, J. "The Second Life of Trees: Family Forestry in Upland Japan." In *The Social Life of Trees,* edited by Laura Rival, 197–218. Oxford: Berg, 1998.

Lynn, C. D. "Hearth and Campfire Influences on Arterial Blood Pressure: Defraying the Costs of the Social Brain Through Fireside Relaxation." *Evolutionary Psychology* 12, no. 5 (2013): 983–1003.

National Printing Bureau (Japan). "Characteristics of Banknotes." 2015. www.npb .go.jp/en/intro/tokutyou/index.html.

Toale, B. *The Art of Papermaking.* Worcester, MA: Davis, 1983.

Vandenbrink, J. P., J. Z. Kiss, R. Herranz, and F. J. Medina. "Light and Gravity Signals Synergize in Modulating Plant Development." *Frontiers in Plant Science* 5 (2014), doi:10.3389/fpls.2014.00563.

Wiessner, P. W. "Embers of Society: Firelight Talk Among the Ju/'hoansi Bushmen." *Proceedings of the National Academy of Sciences* 111, no. 39 (2014): 14027–35.

Woo, S., E. A. Lumpkin, and A. Patapoutian. "Merkel Cells and Neurons Keep in Touch." *Trends in Cell Biology* 25, no. 2 (2015): 74–81.

Wordsworth, W. "A Poet! He Hath Put His Heart to School." 1842. Available at Poetry Foundation, www.poetryfoundation.org/poems-and-poets/poems/detail /45541. Source of "stagnant pool."

———. "The Tables Turned." 1798. Available at Poetry Foundation, www.poetry foundation.org/poems-and-poets/poems/detail/45557. Source of "the beauteous . . ." and "Science and Art."

Ceibo

Araujo, A. "Petroamazonas Perforó el Primer Pozo para Extraer Crudo del ITT." *El Comercio*, March 29, 2016. www.elcomercio.com/actualidad/petroamazonas -perforacion-crudo-yasuniitt.html.

Bass, M. S., M. Finer, C. N. Jenkins, H. Kreft, D. F. Cisneros-Heredia, S. F. McCracken, N. C. A. Pitman, et al. "Global Conservation Significance of Ecuador's Yasuní National Park." *PLoS ONE* 5, no. 1 (2010), doi:10.1371/journal.pone.0008767.

Cerón, C., and C. Montalvo. *Etnobotánica de los Huaorani de Quehueiri-Ono Napo-Ecuador.* Quito: Herbario Alfredo Paredes, Escuela de Biología, Universidad Central del Ecuador, 1998.

Davidson, D. W., S. C. Cook, R. R. Snelling, and T. H. Chua. "Explaining the Abundance of Ants in Lowland Tropical Rainforest Canopies." *Science* 300, no. 5621 (2003): 969–72.

Dillard, A. *Pilgrim at Tinker Creek.* New York: Harper's Magazine Press, 1974. Source of "lifted and struck."

Finer, M., B. Babbitt, S. Novoa, F. Ferrarese, S. Eugenio Pappalardo, M. De Marchi, M. Saucedo, and A. Kumar. "Future of Oil and Gas Development in the Western Amazon." *Environmental Research Letters* 10, no. 2 (2015), doi:10.1088/1748-9326/10/2/024003.

Goffredi, S. K., G. E. Jang, and M. F. Haroon. "Transcriptomics in the Tropics: Total RNA-Based Profiling of Costa Rican Bromeliad-Associated Communities." *Computational and Structural Biotechnology Journal* 13 (2015): 18–23.

Gray, C. L., R. E. Bilsborrow, J. L. Bremner, and F. Lu. "Indigenous Land Use in the Ecuadorian Amazon: A Cross-cultural and Multilevel Analysis." *Human Ecology* 36, no. 1 (2008): 97–109.

Hebdon, C., and F. Mezzenzana. "Sumak Kawsay as 'Already-Developed': A Pastaza Runa Critique of Development." Article draft presented at the Development Studies Association Conference, University of Oxford, September12–14, 2016, Oxford.

Jenkins, C. N., S. L. Pimm, and L. N. Joppa. "Global Patterns of Terrestrial Vertebrate Diversity and Conservation." *Proceedings of the National Academy of Sciences* 110, no. 28 (2013): E2602–10.

Kohn, E. *How Forests Think: Toward an Anthropology Beyond the Human.* Oakland: University of California Press, 2013.

Kursar, T. A., K. G. Dexter, J. Lokvam, R. Toby Pennington, J. E. Richardson, M. G. Weber, E. T. Murakami, C. Drake, R. McGregor, and P. D. Coley. "The Evolution of Antiherbivore Defenses and Their Contribution to Species Coexistence in the Tropical Tree Genus *Inga*." *Proceedings of the National Academy of Sciences* 106, no. 43 (2009): 18073–78.

Lowman, M. D., and H. B. Rinker, eds. *Forest Canopies.* 2nd ed. Burlington, MA: Elsevier, 2004.

McCracken, S. F. and M. R. J. Forstner. "Oil Road Effects on the Anuran Community of a High Canopy Tank Bromeliad (*Aechmea zebrina*) in the Upper Amazon Basin, Ecuador." *PLoS ONE* 9, no. 1 (2014), doi:10.1371/journal.pone.0085470.

Mena, V. P., J. R. Stallings, J. B. Regalado, and R. L. Cueva. "The Sustainability of Current Hunting Practices by the Huaorani." In *Hunting for Sustainability in Tropical Forests*, edited by J. Robinson and E. Bennett, 57–78. New York: Columbia University Press, 2000.

Miroff, N. "Commodity Boom Extracting Increasingly Heavy Toll on Amazon Forests." *Guardian Weekly*, January 9, 2015, pages 12–13.

Nebel, G., L. P. Kvist, J. K. Vanclay, H. Christensen, L. Freitas, and J. Ruíz. "Structure and Floristic Composition of Flood Plain Forests in the Peruvian Amazon: I. Overstorey." *Forest Ecology and Management* 150, no. 1 (2001): 27–57.

Rival, L. "Towards an Understanding of the Huaorani Ways of Knowing and Naming Plants." In *Mobility and Migration in Indigenous Amazonia: Contemporary Ethnoecological Perspectives*, edited by Miguel N. Alexiades, 47–68. New York: Berghahn, 2009.

Rival, L. W. *Trekking Through History: The Huaorani of Amazonian Ecuador*. New York: Columbia University Press, 2002.

Sabagh, L. T., R. J. P. Dias, C. W. C. Branco, and C. F. D. Rocha. "New Records of Phoresy and Hyperphoresy Among Treefrogs, Ostracods, and Ciliates in Bromeliad of Atlantic Forest." *Biodiversity and Conservation* 20, no. 8 (2011): 1837–41.

Schultz, T. R., and S. G. Brady. "Major Evolutionary Transitions in Ant Agriculture." *Proceedings of the National Academy of Sciences* 105, no. 14 (2008): 5435–40.

Suárez, E., M. Morales, R. Cueva, V. Utreras Bucheli, G. Zapata-Ríos, E. Toral, J. Torres, W. Prado, and J. Vargas Olalla. "Oil Industry, Wild Meat Trade and Roads: Indirect Effects of Oil Extraction Activities in a Protected Area in North-Eastern Ecuador." *Animal Conservation* 12, no. 4 (2009): 364–73.

Suárez, E., G. Zapata-Ríos, V. Utreras, S. Strindberg, and J. Vargas. "Controlling Access to Oil Roads Protects Forest Cover, but Not Wildlife Communities: A Case Study from the Rainforest of Yasuní Biosphere Reserve (Ecuador)." *Animal Conservation* 16, no. 3 (2013): 265–74.

Thoreau, H. D. *Walden*. 1854. Available at Digital Thoreau, digitalthoreau.org/fluid-text-toc.

Vidal, J. "Ecuador Rejects Petition to Stop Drilling in National Park." *Guardian Weekly*, May 16, 2014, page 13.

Viteri Gualinga, C. "Visión Indígena del Desarrollo en la Amazonía." *Polis: Revista del Universidad Bolivariano* 3 (2002), doi:10.4000/polis.7678.

Wade, L. "How the Amazon Became a Crucible of Life." *Science*, October 28, 2015. www.sciencemag.org/news/2015/10/feature-how-amazon-became-crucible-life.

Watts, J. "Ecuador Approves Yasuni National Park Oil Drilling in Amazon Rainforest." *Guardian*, August 13, 2013.

Balsam Fir

An, Y. S., B. Kriengwatana, A. E. Newman, E. A. MacDougall-Shackleton, and S. A. MacDougall-Shackleton. "Social Rank, Neophobia and Observational Learning in Black-capped Chickadees." *Behaviour* 148, no. 1 (2011): 55–69.

Aplin, L. M., D. R. Farine, J. Morand-Ferron, A. Cockburn, A. Thornton, and B. C. Sheldon. "Experimentally Induced Innovations Lead to Persistent Culture via Conformity in Wild Birds." *Nature* 518, no. 7540 (2015): 538–41.

Appel, H. M., and R. B. Cocroft. "Plants Respond to Leaf Vibrations Caused by Insect Herbivore Chewing." *Oecologia* 175, no. 4 (2014): 1257–66.

Averill, C., B. L. Turner, and A. C. Finzi. "Mycorrhiza-Mediated Competition Between Plants and Decomposers Drives Soil Carbon Storage." *Nature* 505, no. 7484 (2014): 543–45.

Awramik, S. M., and E. S. Barghoorn. "The Gunflint Microbiota." *Precambrian Research* 5, no. 2 (1977): 121–42.

Babikova, Z., L. Gilbert, T. J. A. Bruce, M. Birkett, J. C. Caulfield, C. Woodcock, J. A. Pickett, and D. Johnson. "Underground Signals Carried Through Common Mycelial Networks Warn Neighbouring Plants of Aphid Attack." *Ecology Letters* 16, no. 7 (2013): 835–43.

Beauregard, P. B., Y. Chai, H. Vlamakis, R. Losick, and R. Kolter. "*Bacillus subtilis* Biofilm Induction by Plant Polysaccharides." *Proceedings of the National Academy of Sciences* 110, no. 17 (2013): E1621–30.

Bond-Lamberty, B., S. D. Peckham, D. E. Ahl, and S. T. Gower. "Fire as the Dominant Driver of Central Canadian Boreal Forest Carbon Balance." *Nature* 450, no. 7166 (2007): 89–92.

Bradshaw, C. J. A., and I. G. Warkentin. "Global Estimates of Boreal Forest Carbon Stocks and Flux." *Global and Planetary Change* 128 (2015): 24–30.

Cossins, D. "Plant Talk." *Scientist* 28, no. 1 (2014): 37–43.

Darwin, C. R. *The Power of Movement in Plants.* London: John Murray, 1880.

Food and Agriculture Organization of the United Nations. *Yearbook of Forest Products.* FAO Forestry Series No. 47, Rome, 2014.

Foote, J. R., D. J. Mennill, L. M. Ratcliffe, and S. M. Smith. "Black-capped Chickadee (*Poecile atricapillus*)." In *The Birds of North America Online*, edited by A. Poole. Ithaca, NY: Cornell Lab of Ornithology, 2010. bna.birds.cornell.edu.bnaproxy .birds.cornell.edu/bna/species/039.

Frederickson, J. K. "Ecological Communities by Design." *Science* 348, no. 6242 (2015): 1425–27.

Ganley, R. J., S. J. Brunsfeld, and G. Newcombe. "A Community of Unknown, Endophytic Fungi in Western White Pine." *Proceedings of the National Academy of Sciences* 101, no. 27 (2004): 10107–12.

Hammerschmidt, K., C. J. Rose, B. Kerr, and P. B. Rainey. "Life Cycles, Fitness Decoupling and the Evolution of Multicellularity." *Nature* 515, no. 7525 (2014): 75–79.

Hansen, M. C., P. V. Potapov, R. Moore, M. Hancher, S. A. Turubanova, A. Tyukavina, D. Thau, et al. "High-Resolution Global Maps of 21st-Century Forest Cover Change." *Science* 342, no. 6160 (2013): 850–53.

Hata, K., and K. Futai. "Variation in Fungal Endophyte Populations in Needles of the Genus *Pinus.*" *Canadian Journal of Botany* 74, no. 1 (1996): 103–14.

Hom, E. F. Y., and A. W. Murray. "Niche Engineering Demonstrates a Latent Capacity for Fungal-Algal Mutualism." *Science* 345, no. 6192 (2014): 94–98.

Hordijk, W. "Autocatalytic Sets: From the Origin of Life to the Economy." *BioScience* 63, no. 11 (2013): 877–81.

Karhu, K., M. D. Auffret, J. A. J. Dungait, D. W. Hopkins, J. I. Prosser, B. K. Singh, J.-A. Subke, et al. "Temperature Sensitivity of Soil Respiration Rates Enhanced by Microbial Community Response." *Nature* 513, no. 7516 (2014): 81–84.

Karzbrun, E., A. M. Tayar, V. Noireaux, and R. H. Bar-Ziv. "Programmable On-Chip DNA Compartments as Artificial Cells." *Science* 345, no. 6198 (2014): 829–32.

Keller, M. A., A. V. Turchyn, and M. Ralser. "Non-enzymatic Glycolysis and Pentose Phosphate Pathway-like Reactions in a Plausible Archean Ocean." *Molecular Systems Biology* 10, no. 4 (2014), doi:10.1002/msb.20145228.

Knoll, A. H., E. S. Barghoorn, and S. M. Awramik. "New Microorganisms from the Aphebian Gunflint Iron Formation, Ontario." *Journal of Paleontology* 52, no. 5 (1978): 976–92.

Libby, E., and W. C. Ratcliff. "Ratcheting the Evolution of Multicellularity." *Science* 346, no. 6208 (2014): 426–27.

Liu, C., T. Liu, F. Yuan, and Y. Gu. "Isolating Endophytic Fungi from Evergreen Plants and Determining Their Antifungal Activities." *African Journal of Microbiology Research* 4, no. 21 (2010): 2243–48.

Lyons, T. W., C. T. Reinhard, and N. J. Planavsky. "The Rise of Oxygen in Earth's Early Ocean and Atmosphere." *Nature* 506, no. 7488 (2014): 307–15.

Molinier, J., G. Ries, C. Zipfel, and B. Hohn. "Transgeneration Memory of Stress in Plants." *Nature* 442, no. 7106 (2006): 1046–49.

Mousavi, S. A. R., A. Chauvin, F. Pascaud, S. Kellenberger, and E. E. Farmer. "Glutamate Receptor-like Genes Mediate Leaf-to-Leaf Wound Signalling." *Nature* 500, no. 7463 (2013): 422–26.

Nelson-Sathi, S., F. L. Sousa, M. Roettger, N. Lozada-Chávez, T. Thiergart, A. Janssen, D. Bryant, et al. "Origins of Major Archaeal Clades Correspond to Gene Acquisitions from Bacteria." *Nature* 517, no. 7532 (2014): 77–80.

Ortiz-Castro, R., C. Díaz-Pérez, M. Martínez-Trujillo, E. Rosa, J. Campos-García, and J. López-Bucio. "Transkingdom Signaling Based on Bacterial Cyclodipeptides with Auxin Activity in Plants." *Proceedings of the National Academy of Sciences* 108, no. 17 (2011): 7253–58.

Pagès, A., K. Grice, M. Vacher, D. T. Welsh, P. R. Teasdale, W. W. Bennett, and P. Greenwood. "Characterizing Microbial Communities and Processes in a Modern Stromatolite (Shark Bay) Using Lipid Biomarkers and Two-Dimensional Distributions of Porewater Solutes." *Environmental Microbiology* 16, no. 8 (2014): 2458–74.

Parniske, M. "Arbuscular Mycorrhiza: The Mother of Plant Root Endosymbioses." *Nature Reviews Microbiology* 6 (2008): 763–75.

Roth, T. C., and V. V. Pravosudov. "Hippocampal Volumes and Neuron Numbers Increase Along a Gradient of Environmental Harshness: A Large-Scale Comparison." *Proceedings of the Royal Society B: Biological Sciences* 276, no. 1656 (2009): 401–5.

Schopf, J. W. "Solution to Darwin's Dilemma: Discovery of the Missing Precambrian Record of Life." *Proceedings of the National Academy of Sciences* 97, no. 13 (2000): 6947–53.

Song, Y. Y., R. S. Zeng, J. F. Xu, J. Li, X. Shen, and W. G. Yihdego. "Interplant Communication of Tomato Plants Through Underground Common Mycorrhizal Networks." *PLoS ONE* 5, no. 10 (2010): e13324.

Stal, L. J. "Cyanobacterial Mats and Stromatolites." In *Ecology of Cyanobacteria II*, edited by B. A. Whitton, 61–120. Dordrecht, Netherlands: Springer, 2012.

Tedersoo, L., T. W. May, and M. E. Smith. "Ectomycorrhizal Lifestyle in Fungi: Global Diversity, Distribution, and Evolution of Phylogenetic Lineages." *Mycorrhiza* 20, no. 4 (2010): 217–63.

Templeton, C. N., and E. Greene. "Nuthatches Eavesdrop on Variations in Heterospecific Chickadee Mobbing Alarm Calls." *Proceedings of the National Academy of Sciences* 104, no. 13 (2007): 5479–82.

Trewavas, A. *Plant Behaviour and Intelligence.* Oxford: Oxford University Press, 2014.

———. "What Is Plant Behaviour?" *Plant, Cell & Environment* 32, no. 6 (2009): 606–16.

Vaidya, N., M. L. Manapat, I. A. Chen, R. Xulvi-Brunet, E. J. Hayden, and N. Lehman. "Spontaneous Network Formation Among Cooperative RNA Replicators." *Nature* 491, no. 7422 (2012): 72–77.

Wacey, D., N. McLoughlin, M. R. Kilburn, M. Saunders, J. B. Cliff, C. Kong, M. E. Barley, and M. D. Brasier. "Nanoscale Analysis of Pyritized Microfossils Reveals Differential Heterotrophic Consumption in the ~1.9-Ga Gunflint Chert." *Proceedings of the National Academy of Sciences* 110, no. 20 (2013): 8020–24.

Woolf, V. *A Room of One's Own.* London: Hogarth Press, 1929.

Sabal Palm

Amin, S. A., L. R. Hmelo, H. M. van Tol, B. P. Durham, L. T. Carlson, K. R. Heal, R. L. Morales, et al. "Interaction and Signaling Between a Cosmopolitan Phytoplankton and Associated Bacteria." *Nature* 522, no. 7554 (2015): 98–101.

Anelay, J. 2014. Written Answers: Mediterranean Sea. October 15, 2014. *Hansard Parliamentary Debates*, Lords, vol. 756, part 39, col. WA41. Source of "We do not support planned search and rescue . . ."

Böhm, E., J. Lippold, M. Gutjahr, M. Frank, P. Blaser, B. Antz, J. Fohlmeister, N. Frank, M. B. Andersen, and M. Deininger. "Strong and Deep Atlantic Meridional Overturning Circulation During the Last Glacial Cycle." *Nature* 517, no. 7532 (2015): 73–76.

Boyce, D. G., M. R. Lewis, and B. Worm. "Global Phytoplankton Decline over the Past Century." *Nature* 466, no. 7306 (2010): 591–96.

Buckley, F. "Thoreau and the Irish." *New England Quarterly* 13, no. 3 (September 1, 1940): 389–400.

Chen, X., and K.-K. Tung. "Varying Planetary Heat Sink Led to Global-Warming Slowdown and Acceleration." *Science* 345, no. 6199 (2014): 897–903.

Cózar, A., F. Echevarría, J. I. González-Gordillo, X. Irigoien, B. Úbeda, S. Hernández-León, Á. T. Palma, et al. "Plastic Debris in the Open Ocean." *Proceedings of the National Academy of Sciences* 111, no. 28 (2014): 10239–44.

Desantis, L. R. G., S. Bhotika, K. Williams, and F. E. Putz. "Sea-Level Rise and Drought Interactions Accelerate Forest Decline on the Gulf Coast of Florida, USA." *Global Change Biology* 13, no. 11 (2007): 2349–60.

Gemenne, F. "Why the Numbers Don't Add Up: A Review of Estimates and Predictions of People Displaced by Environmental Changes." *Global Environmental Change* 21 (2011): S41–49.

Gráda, C. O. "A Note on Nineteenth-Century Irish Emigration Statistics." *Population Studies* 29, no. 1 (1975): 143–49.

Hay, C. C., E. Morrow, R. E. Kopp, and J. X. Mitrovica. "Probabilistic Reanalysis of Twentieth-Century Sea-Level Rise." *Nature* 517, no. 7535 (2015): 481–84.

Holbrook, N. M., and T. R. Sinclair. "Water Balance in the Arborescent Palm, *Sabal palmetto*. I. Stem Structure, Tissue Water Release Properties and Leaf Epidermal Conductance." *Plant, Cell & Environment* 15, no. 4 (1992): 393–99.

———. "Water Balance in the Arborescent Palm, *Sabal palmetto*. II. Transpiration and Stem Water Storage." *Plant, Cell & Environment* 15, no. 4 (1992): 401–9.

Jambeck, J. R., R. Geyer, C. Wilcox, T. R. Siegler, M. Perryman, A. Andrady, R. Narayan, and K. L. Law. "Plastic Waste Inputs from Land into the Ocean." *Science* 347, no. 6223 (2015): 768–71.

Joughin, I., B. E. Smith, and B. Medley. "Marine Ice Sheet Collapse Potentially Under Way for the Thwaites Glacier Basin, West Antarctica." *Science* 344, no. 6185 (2014): 735–38.

Lee, D. S. "Floridian Herpetofauna Associated with Cabbage Palms." *Herpetologica* 25 (1969): 70–71.

Limardo, A. J., and A. Z. Worden. "Microbiology: Exclusive Networks in the Sea." *Nature* 522, no. 7554 (2015): 36–37.

Mansfield, K. L., J. Wyneken, W. P. Porter, and J. Luo. "First Satellite Tracks of Neonate Sea Turtles Redefine the 'Lost Years' Oceanic Niche." *Proceedings of the Royal Society B: Biological Sciences* 281, no. 1781 (2014), doi:10.1098/rspb.2013.3039.

Maranger, R., and D. F. Bird. "Viral Abundance in Aquatic Systems: A Comparison Between Marine and Fresh Waters." *Marine Ecology Progress Series* 121 (1995): 217–26.

McPherson, K., and K. Williams. "Establishment Growth of Cabbage Palm, *Sabal palmetto* (Arecaceae)." *American Journal of Botany* 83, no. 12 (1996): 1566–70.

———. "The Role of Carbohydrate Reserves in the Growth, Resilience, and Persistence of Cabbage Palm Seedlings (*Sabal palmetto*)." *Oecologia* 117, no. 4 (1998): 460–68.

Meyer, B. K., G. A. Bishop, and R. K. Vance. "An Evaluation of Shoreline Dynamics at St. Catherine's Island, Georgia (1859–2009) Utilizing the Digital Shoreline Analysis System (USGS)." *Geological Society of America Abstracts with Programs* 43, no. 2 (2011): 68.

Morris, J. J., R. E. Lenski, and E. R. Zinser. "The Black Queen Hypothesis: Evolution of Dependencies Through Adaptive Gene Loss." *MBio* 3, no. 2 (2012), doi:10.1128/mBio.00036-12.

National Park Service. "Cape Cod National Seashore: Shipwrecks." N.d. www.nps.gov/caco/learn/historyculture/shipwrecks.htm (accessed May 7, 2015).

Nicholls, R. J., N. Marinova, J. A. Lowe, S. Brown, P. Vellinga, D. De Gusmao, J. Hinkel, and R. S. J. Tol. "Sea-Level Rise and Its Possible Impacts Given a 'Beyond 4 C World' in the Twenty-first Century." *Philosophical Transactions of the Royal Society A: Mathematical, Physical and Engineering Sciences* 369, no. 1934 (2011): 161–81.

Nuwer, R. "Plastic on Ice." *Scientific American* 311, no. 3 (2014): 25.

Osborn, A. M., and S. Stojkovic. "Marine Microbes in the Plastic Age." *Microbiology Australia* 35, no. 4 (2014): 207–10.

Paolo, F. S., H. A. Fricker, and L. Padman. "Volume Loss from Antarctic Ice Is Accelerating." *Science* 348 (2015): 327–31.

Perry, L., and K. Williams. "Effects of Salinity and Flooding on Seedlings of Cabbage Palm (*Sabal palmetto*)." *Oecologia* 105, no. 4 (1996): 428–34.

Reisser, J., B. Slat, K. Noble, K. du Plessis, M. Epp, M. Proietti, J. de Sonneville, T. Becker, and C. Pattiaratchi. "The Vertical Distribution of Buoyant Plastics at Sea: An Observational Study in the North Atlantic Gyre." *Biogeosciences* 12, no. 4 (2015): 1249–56.

Rohling, E. J., G. L. Foster, K. M. Grant, G. Marino, A. P. Roberts, M. E. Tamisiea, and F. Williams. "Sea-Level and Deep-Sea-Temperature Variability over the Past 5.3 Million Years." *Nature* 508, no. 7497 (2014): 477–82.

Swan, B. K., B. Tupper, A. Sczyrba, F. M. Lauro, M. Martinez-Garcia, J. M. González, H. Luo, et al. "Prevalent Genome Streamlining and Latitudinal Divergence of Planktonic Bacteria in the Surface Ocean." *Proceedings of the National Academy of Sciences* 110, no. 28 (2013): 11463–68.

Thomas, D. H., C. F. T. Andrus, G. A. Bishop, E. Blair, D. B. Blanton, D. E. Crowe, C. B. DePratter, et al. "Native American Landscapes of St. Catherines Island, Georgia." *Anthropological Papers of the American Museum of Natural History*, no. 88 (2008).

Thoreau, H. D. *Cape Cod.* Boston: Ticknor and Fields, 1865. Source of "waste and wrecks . . . ," "why waste . . . ," and quotes from beach list.

Tomlinson P. B. "The Uniqueness of Palms." *Botanical Journal of the Linnean Society* 151 (2006): 5–14.

Tomlinson, P. B., J. W. Horn, and J. B. Fisher. *The Anatomy of Palms.* Oxford: Oxford University Press, 2011.

U.S. Department of Defense. *FY 2014 Climate Change Adaptation Roadmap.* Alexandria, VA: Office of the Deputy Undersecretary of Defense for Installations and Environment, 2014.

Woodruff, J. D., J. L. Irish, and S. J. Camargo. "Coastal Flooding by Tropical Cyclones and Sea-Level Rise." *Nature* 504, no. 7478 (2013): 44–52.

Wright, S. L., D. Rowe, R. C. Thompson, and T. S. Galloway. "Microplastic Ingestion Decreases Energy Reserves in Marine Worms." *Current Biology* 23, no. 23 (2013): R1031–33.

Zettler, E. R., T. J. Mincer, and L. A. Amaral-Zettler. "Life in the 'Plastisphere': Microbial Communities on Plastic Marine Debris." *Environmental Science & Technology* 47, no. 13 (2013): 7137–46.

Zona, S. "A Monograph of *Sabal* (Arecaceae: Coryphoideae)." *Aliso* 12, no. 4 (1990): 583–666.

Green Ash

Allender, M. C., D. B. Raudabaugh, F. H. Gleason, and A. N. Miller. "The Natural History, Ecology, and Epidemiology of *Ophidiomyces ophiodiicola* and Its Potential Impact on Free-Ranging Snake Populations." *Fungal Ecology* 17 (2015): 187–96.

Chambers, J. Q., N. Higuchi, J. P. Schimel, L. V. Ferreira, and J. M. Melack. "Decomposition and Carbon Cycling of Dead Trees in Tropical Forests of the Central Amazon." *Oecologia* 122, no. 3 (2000): 380–88.

Gerdeman, B. S., and G. Rufino. "Heterozerconidae: A Comparison Between a Temperate and a Tropical Species." In *Trends in Acarology, Proceedings of the 12th International Congress*, edited by M. W. Sabelis and J. Bruin, 93–96. Dordrecht, Netherlands: Springer, 2011.

Hérault, B., J. Beauchêne, F. Muller, F. Wagner, C. Baraloto, L. Blanc, and J. Martin. "Modeling Decay Rates of Dead Wood in a Neotropical Forest." *Oecologia* 164, no. 1 (2010): 243–51.

Hulcr, J., N. R. Rountree, S. E. Diamond, L. L. Stelinski, N. Fierer, and R. R. Dunn. "Mycangia of Ambrosia Beetles Host Communities of Bacteria." *Microbial Ecology* 64, no. 3 (2012): 784–93.

Pan, Y., R. A. Birdsey, J. Fang, R. Houghton, P. E. Kauppi, W. A. Kurz, O. L. Phillips, et al. "A Large and Persistent Carbon Sink in the World's Forests." *Science* 333, no. 6045 (2011): 988–93.

Rodrigues, R. R., R. P. Pineda, J. N. Barney, E. T. Nilsen, J. E. Barrett, and M. A. Williams. "Plant Invasions Associated with Change in Root-Zone Microbial Community Structure and Diversity." *PLoS ONE* 10, no. 10 (2015): e0141424.

Vandenbrink, J. P., J. Z. Kiss, R. Herranz, and F. J. Medina. "Light and Gravity Signals Synergize in Modulating Plant Development." *Frontiers in Plant Science* 5 (2014), doi:10.3389/fpls.2014.00563.

Hazel

BBC Radio 4. Interviews of Dorothy Thompson, CEO Drax Group, and Harry Huyton, Head of Climate Change Policy and Campaigns, RSPB. *Today*, July 24, 2014.

Birks, H. J. B. "Holocene Isochrone Maps and Patterns of Tree-Spreading in the British Isles." *Journal of Biogeography* 16, no. 6 (1989): 503–40.

Bishop, R. R., M. J. Church, and P. A. Rowley-Conwy. "Firewood, Food and Human Niche Construction: The Potential Role of Mesolithic Hunter-Gatherers in Actively Structuring Scotland's Woodlands." *Quaternary Science Reviews* 108 (2015): 51–75.

Carlyle, T. *Historical Sketches of Notable Persons and Events in the Reigns of James I and Charles I*. London: Chapman and Hall, 1898.

Carrell, S. "Longannet Power Station to Close Next Year." *Guardian*, March 23, 2015.

Climate Change (Scotland) Act 2009. www.legislation.gov.uk/asp/2009/12/contents (accessed June 1, 2015).

Dinnis, R., and C. Stringer. *Britain: One Million Years of the Human Story*. London: Natural History Museum Publications, 2014.

Edwards, K. J., and I. Ralston. "Postglacial Hunter-Gatherers and Vegetational History in Scotland." *Proceedings of the Society of Antiquaries of Scotland* 114 (1984): 15–34.

Evans, J. M., R. J. Fletcher Jr., J. R. R. Alavalapati, A. L. Smith, D. Geller, P. Lal, D. Vasudev, M. Acevedo, J. Calabria, and T. Upadhyay. *Forestry Bioenergy in the Southeast United States: Implications for Wildlife Habitat and Biodiversity*. Merrifield, VA: National Wildlife Federation, 2013.

Finsinger, W., W. Tinner, W. O. Van der Knaap, and B. Ammann. "The Expansion of Hazel (*Corylus avellana* L.) in the Southern Alps: A Key for Understanding Its Early Holocene History in Europe?" *Quaternary Science Reviews* 25, no. 5 (2006): 612–31.

Fodor, E. "Linking Biodiversity to Mutualistic Networks: Woody Species and Ectomycorrhizal Fungi." *Annals of Forest Research* 56 (2012): 53–78.

Furniture Industry Research Association. "Biomass Subsidies and Their Impact on the British Furniture Industry." Stevenage, UK, 2011.

Glasgow Herald. "Scots Pit Props: Developing a Rural Industry," January 8, 1938, page 3.

Mather, A. S. "Forest Transition Theory and the Reforesting of Scotland." *Scottish Geographical Magazine* 120, no. 1–2 (2004): 83–98.

Meyfroidt, P., T. K. Rudel, and E. F. Lambin. "Forest Transitions, Trade, and the Global Displacement of Land Use." *Proceedings of the National Academy of Sciences* 107, no. 49 (2010): 20917–22.

Palmé, A. E., and G. C. Vendramin. "Chloroplast DNA Variation, Postglacial Recolonization and Hybridization in Hazel, *Corylus avellana*." *Molecular Ecology* 11 (2002): 1769–79.

Regnell, M. "Plant Subsistence and Environment at the Mesolithic Site Tågerup, Southern Sweden: New Insights on the 'Nut Age.'" *Vegetation History and Archaeobotany* 21 (2012): 1–16.

Robertson, A., J. Lochrie, and S. Timpany. "Built to Last: Mesolithic and Neolithic Settlement at Two Sites Beside the Forth Estuary, Scotland." *Proceedings of the Society of Antiquaries of Scotland* 143 (2013): 1–64.

Schoch, W., I. Heller, F. H. Schweingruber, and F. Kienast. "Wood Anatomy of Central European Species." 2004. www.woodanatomy.ch.

Scott, W. *The Abbot*. Edinburgh: Longman, 1820.

Scottish Government. "High Level Summary of Statistics Trend Last Update: Renewable Energy. December 18, 2014. www.gov.scot/Topics/Statistics/Browse/Business/TrenRenEnergy.

Scottish Mining. "Accidents and Disasters." www.scottishmining.co.uk/5.html.

Soden, L. 2012. *Landscape Management Plan*. Rosyth, UK: Forth Crossing Bridge Constructors, 2012. www.transport.gov.scot/system/files/documents/tsc-basic-pages/10%20REP-00028-01%20Landscape%20Management%20Plan%20%28EM%20update%20for%20website%29.pdf.

Stephenson, A. L., and D. J. C. MacKay. *Life Cycle Impacts of Biomass Electricity in 2020: Scenarios for Assessing the Greenhouse Gas Impacts and Energy Input Requirements of Using North American Woody Biomass for Electricity Generation in the UK.* London: United Kingdom Department of Energy and Climate Change, 2014.

Stevenson, R. L. *Kidnapped.* New York and London: Harper, 1886.

"The Supply of Pitwood." *Nature* 94 (1914): 393–95.

Tallantire, P. A. "The Early-Holocene Spread of Hazel (*Corylus avellana* L.) in Europe North and West of the Alps: An Ecological Hypothesis." *Holocene* 12 (2002): 81–96.

Ter-Mikaelian, M. T., S. J. Colombo, and J. Chen. "The Burning Question: Does Forest Bioenergy Reduce Carbon Emissions? A Review of Common Misconceptions About Forest Carbon Accounting." *Journal of Forestry* 113, no. 1 (2015): 57–68.

United Kingdom. *Electricity, England and Wales: Renewables Obligation Order 2009.* Statutory Instrument 2009/785, March 24, 2009.

———. Office of Gas and Electricity Markets. "Renewables Obligation (RO) Annual Report 2013–14." February 16, 2015. www.ofgem.gov.uk//publications-and-updates/renewables-obligation-ro-annual-report-2013-14.

U.S. Energy Information Administration. *International Energy Statistics.* Washington, DC: U.S. Department of Energy, 2015. www.eia.gov/beta/international/.

U.S. Environmental Protection Agency. *Framework for Assessing Biogenic CO_2 Emissions from Stationary Sources.* Washington, DC: Office of Air and Radiation, Office of Atmospheric Programs, Climate Change Division, 2014.

West Fife Council. 1994. "Kingdom of Fife Mining Industry Memorial Book." www.fifepits.co.uk/starter/m-book.htm/.

Warrick, J. 2015. "How Europe's Climate Policies Led to More U.S. Trees Being Cut Down." *Washington Post*, June 2, 2105. wpo.st/bARKo.

Redwood and Ponderosa Pine

Allen, C. D., A. K. Macalady, H. Chenchouni, D. Bachelet, N. McDowell, M. Vennetier, T. Kitzberger, et al. "A Global Overview of Drought and Heat-Induced Tree Mortality Reveals Emerging Climate Change Risks for Forests." *Forest Ecology and Management* 259, no. 4 (2010): 660–84.

Baker, J. A. *The Peregrine.* London: Collins, 1967.

Bannan, M. W. "The Length, Tangential Diameter, and Length/Width Ratio of Conifer Tracheids." *Canadian Journal of Botany* 43, no. 8 (1965): 967–84.

Bijl, P. K., A. J. P. Houben, S. Schouten, S. M. Bohaty, A. Sluijs, G.-J. Reichart, J. S. Sinninghe Damsté, and H. Brinkhuis. "Transient Middle Eocene Atmospheric CO_2 and Temperature Variations." *Science* 330, no. 6005 (2010), doi:10.1126/science.1193654.

Borsa, A. A., D. C. Agnew, and D. R. Cayan. "Ongoing Drought-Induced Uplift in the Western United States." *Science* 345, no. 6204 (2014), doi:10.1126/science.1260279.

Callaham, R. Z. "*Pinus ponderosa*: Geographic Races and Subspecies Based on Morphological Variation." Research Paper PSW-RP-265, U.S. Department of Agriculture, Forest Service, Pacific Southwest Research Station, Albany, CA, 2013.

Carswell, C. "Don't Blame the Beetles." *Science* 346, no. 6206 (2014), doi:10.1126/science.346.6206.154.

Chapman, S. S., G. E. Griffith, J. M. Omernik, A. B. Price, J. Freeouf, and D. L. Schrupp. *Ecoregions of Colorado.* Reston, VA: U.S. Geological Survey, 2006.

DeConto, R. M., and D. Pollard. "Rapid Cenozoic Glaciation of Antarctica Induced by Declining Atmospheric CO_2." *Nature* 421, no. 6920 (2003): 245–49.

Domec, J. C., J. M. Warren, F. C. Meinzer, J. R. Brooks, and R. Coulombe. "Native Root Xylem Embolism and Stomatal Closure in Stands of Douglas-Fir and Ponderosa Pine: Mitigation by Hydraulic Redistribution." *Oecologia* 141, no. 1 (2004): 7–16.

Editorial Board. "Congress Should Give the Government More Money for Wildfires." *New York Times*, September 28, 2015. www.nytimes.com/2015/09/28/opinion/congress-should-give-the-government-more-money-for-wildfires.html.

Evanoff, E., K. M. Gregory-Wodzicki, and K. R. Johnson, eds. *Fossil Flora and Stratigraphy of the Florissant Formation, Colorado.* Denver: Denver Museum of Nature and Science, 2011.

Feynman, R. *The Character of Physical Law.* Cambridge: MIT Press, 1967. Source of "nature has a simplicity" and "the deepest beauty."

Frost, R. "The Sound of Trees." *The Poetry of Robert Frost: The Collected Poems, Complete and Unabridged.* New York: Holt, 2002. Source of "all measure . . ."

Ganey, J. L., and S. C. Vojta. "Tree Mortality in Drought-Stressed Mixed-Conifer and Ponderosa Pine Forests, Arizona, USA." *Forest Ecology and Management* 261, no. 1 (2011): 162–68.

Hume, D. *Four Dissertations. IV. Of the Standard of Taste.* 1757. Available at www.davidhume.org/texts/fd.html. Source of "Beauty is no quality in things . . ." and "Strong sense, united to delicate sentiment . . ."

Kawabata, Y. *Snow Country.* Translated by E. G. Seidensticker. New York: A. A. Knopf, 1956.

Keegan, K. M., M. R. Albert, J. R. McConnell, and I. Baker. "Climate Change and Forest Fires Synergistically Drive Widespread Melt Events of the Greenland Ice Sheet." *Proceedings of the National Academy of Sciences* 111, no. 22 (2014), doi:10.1073/pnas.1405397111.

Keller, L., and M. G. Surette. "Communication in Bacteria: An Ecological and Evolutionary Perspective." *Nature Reviews Microbiology* 4, no. 4 (2006): 249–58.

Kikuta, S. B., M. A. Lo Gullo, A. Nardini, H. Richter, and S. Salleo. "Ultrasound Acoustic Emissions from Dehydrating Leaves of Deciduous and Evergreen Trees." *Plant, Cell & Environment* 20, no. 11 (1997): 1381–90.

Laschimke, R., M. Burger, and H. Vallen. "Acoustic Emission Analysis and Experiments with Physical Model Systems Reveal a Peculiar Nature of the Xylem Tension." *Journal of Plant Physiology* 163, no. 10 (2006): 996–1007.

Maherali, H., and E. H. DeLucia. "Xylem Conductivity and Vulnerability to Cavitation of Ponderosa Pine Growing in Contrasting Climates." *Tree Physiology* 20, no. 13 (2000): 859–67.

Maxbauer, D. P., D. L. Royer, and B. A. LePage. "High Arctic Forests During the Middle Eocene Supported by Moderate Levels of Atmospheric CO_2." *Geology* 42, no. 12 (2014): 1027–30.

Meko, D. M., C. A. Woodhouse, C. A. Baisan, T. Knight, J. J. Lukas, M. K. Hughes, and M. W. Salzer. "Medieval Drought in the Upper Colorado River Basin." *Geophysical Research Letters* 34, no. 10 (2007), doi:10.1029/2007GL029988.

Meyer, H. W. *The Fossils of Florissant.* Washington, DC: Smithsonian Books, 2003.

Monson, R. K., and M. C. Grant. "Experimental Studies of Ponderosa Pine. III. Differences in Photosynthesis, Stomatal Conductance, and Water-Use Efficiency Between Two Genetic Lines." *American Journal of Botany* 76, no. 7 (1989): 1041–47.

Moritz, M. A., E. Batllori, R. A. Bradstock, A. M. Gill, J. Handmer, P. F. Hessburg, J. Leonard, et al. "Learning to Coexist with Wildfire." *Nature* 515, no. 7525 (2014), doi:10.1038/nature13946.

Muir, J. *The Mountains of California.* New York: Century Company, 1894. Source of "finest music . . . hum."

Murdoch, I. *The Sovereignty of Good.* London: Routledge, 1970. Source of "unselfing" and "patently . . ."

Oliver, W. W., and R. A. Ryker. "Ponderosa Pine." In *Silvics of North America*, edited by R. M. Burns and B. H. Honkala. Agriculture Handbook 654. U.S. Department of Agriculture, Forest Service, Washington, DC, 1990. www.na.fs.fed.us /spfo/pubs/silvics_manual/Volume_1/pinus/ponderosa.htm.

Pais, A., M. Jacob, D. I. Olive, and M. F. Atiyah. *Paul Dirac: The Man and His Work.* Cambridge, UK: Cambridge University Press, 1998. Source of "getting beauty . . ."

Pierce, J. L., G. A. Meyer, and A. J. T. Jull. "Fire-Induced Erosion and Millennial-Scale Climate Change in Northern Ponderosa Pine Forests." *Nature* 432, no. 7013 (2004), doi:10.1038/nature03058.

Pross, J., L. Contreras, P. K. Bijl, D. R. Greenwood, S. M. Bohaty, S. Schouten, J. A. Bendle, et al. "Persistent Near-Tropical Warmth on the Antarctic Continent During the Early Eocene Epoch." *Nature* 488, no. 7409 (2012), doi:10.1038/nature11300.

Ryan, M. G., B. J. Bond, B. E. Law, R. M. Hubbard, D. Woodruff, E. Cienciala, and J. Kucera. "Transpiration and Whole-Tree Conductance in Ponderosa Pine Trees of Different Heights." *Oecologia* 124, no. 4 (2000): 553–60.

Shen, F., Y. Wang, Y. Cheng, and L. Zhang. "Three Types of Cavitation Caused by Air Seeding." *Tree Physiology* 32, no. 11 (2012): 1413–19.

Svensen, H., S. Planke, A. Malthe-Sørenssen, B. Jamtveit, R. Myklebust, T. R. Eidem, and S. S. Rey. "Release of Methane from a Volcanic Basin as a Mechanism for Initial Eocene Global Warming." *Nature* 429, no. 6991 (2004), doi:10.1038 /nature02566.

Underwood, E. "Models Predict Longer, Deeper U.S. Droughts." *Science* 347, no. 6223 (2015), doi:10.1126/science.347.6223.707. Source of "quaint."

van Riper III, C., J. R. Hatten, J. T. Giermakowski, D. Mattson, J. A. Holmes, M. J. Johnson, E. M. Nowak, et al. "Projecting Climate Effects on Birds and Reptiles of the Southwestern United States." U.S. Geological Survey Open-File Report 2014-1050, 2014, doi:10.3133/ofr20141050.

Warren, J. M., J. R. Brooks, F. C. Meinzer, and J. L. Eberhart. "Hydraulic Redistribution of Water from *Pinus ponderosa* Trees to Seedlings: Evidence for an Ectomycorrhizal Pathway." *New Phytologist* 178, no. 2 (2008): 382–94.

Weed, A. S., M. P. Ayres, and J. A. Hicke. "Consequences of Climate Change for Biotic Disturbances in North American Forests." *Ecological Monographs* 83, no. 4 (2013): 441–70.

Westerling, A. L., H. G. Hidalgo, D. R. Cayan, and T. W. Swetnam. "Warming and Earlier Spring Increase Western US Forest Wildfire Activity." *Science* 313, no. 5789 (2006): 940–43.

Zachos, J., M. Pagani, L. Sloan, E. Thomas, and K. Billups. "Trends, Rhythms, and Aberrations in Global Climate 65 Ma to Present." *Science* 292, no. 5517 (2001): 686–93.

Zhang, Y. G., M. Pagani, Z. Liu, S. M. Bohaty, and R. DeConto. (2013). "A 40-Million-Year History of Atmospheric CO_2." *Philosophical Transactions of the Royal Society A: Mathematical, Physical and Engineering Sciences* 371, no. 2001 (2013), doi:10.1098/rsta.2013.0096.

Cottonwood

Barbaccia, T. G. "A Benchmark for Snow and Ice Management in the Mile High City." *Equipment World's Better Roads*, August 25, 2010. www.equipmentworld.com/a-benchmark-for-snow-and-ice-management-in-the-mile-high-city/.

Belk, J. 2003. "Big Sky, Open Arms." *New York Times*, June 22, 2003. www.nytimes.com/2003/06/22/travel/big-sky-open-arms.html. Source of "Four black folks . . ."

Blasius, B. J., and R. W. Merritt. "Field and Laboratory Investigations on the Effects of Road Salt (NaCl) on Stream Macroinvertebrate Communities." *Environmental Pollution* 120, no. 2 (2002): 219–31.

Clancy, K. B. H., R. G. Nelson, J. N. Rutherford, and K. Hinde. "Survey of Academic Field Experiences (SAFE): Trainees Report Harassment and Assault." *PLoS ONE* 9, no. 7 (July 16, 2014), doi:10.1371/journal.pone.0102172. Source of "hostile field environments."

Coates, T. *Between the World and Me*. New York: Spiegel & Grau, 2015. Source of "Catholic, Corsican . . ."

Conathan, L., ed. "Arapaho text corpus." Endangered Language Archive, 2006. elar.soas.ac.uk/deposit/0083.

Davidson, J. "Former Legislator Joe Shoemaker Led Cleanup of the S. Platte River." *Denver Post*, August 16, 2012. www.denverpost.com/ci_21323273/former-legislator-joe-shoemaker-led-cleanup-s-platte.

Dillard, A. "Innocence in the Galapagos." *Harper's*, May 1975. Source of "pristine . . ." and "the greeting . . ."

Finney, C. *Black Faces, White Spaces: Reimagining the Relationship of African Americans to the Great Outdoors*. Chapel Hill: University of North Carolina Press, 2014. Source of "geographies of fear."

Greenway Foundation. *The River South Greenway Master Plan.* Greenwood Village, CO: Greenway Foundation, 2010. www.thegreenwayfoundation.org/uploads/3 /9/1/5/39157543/riso.pdf.

———. *The Greenway Foundation Annual Report.* Denver, CO: Greenway Foundation, April 2012. www.thegreenwayfoundation.org/uploads/3/9/1/5/39157543/2012 _greenway_current.pdf.

Gwaltney, B. Interviewed in "James Mills on African Americans and National Parks." To the Best of Our Knowledge, August 29, 2010. www.ttbook.org/book/james-mills -african-americans-and-national-parks. Source of "There are a lot of trees . . ."

Jefferson, T. "Notes on the State of Virginia." 1787. Available at Yale University Avalon Project. avalon.law.yale.edu/18th_century/jeffvir.asp. Source of "mobs of great cities . . ." and "husbandmen."

Kranjcec, J., J. M. Mahoney, and S. B. Rood. "The Responses of Three Riparian Cottonwood Species to Water Table Decline." *Forest Ecology and Management* 110, no. 1 (1998): 77–87.

Lanham, J. D. "9 Rules for the Black Birdwatcher." *Orion* 32, no. 6 (November 1, 2013): 7. Source of "Don't bird . . ."

Leopold, A. "The Last Stand of the Wilderness." *American Forests and Forest Life* 31, no. 382 (October 1925): 599–604. Source of "segregated . . . wilderness."

———. *A Sand County Almanac, and Sketches Here and There.* Oxford: Oxford University Press, 1949. Source of "soils, waters . . ." and "man-made changes."

Limerick, P. N. *A Ditch in Time: The City, the West, and Water.* Golden, CO: Fulcrum, 2012. Source of "perpetually brilliant" and "tonic, healthy."

Louv. R. *Last Child in the Woods.* Chapel Hill, NC: Algonquin, 2005. Source of "nature deficit."

Marotti, A. "Denver's Camping Ban: Survey Says Police Don't Help Homeless Enough." *Denver Post,* June 26, 2013. www.denverpost.com/politics/ci_23539228/denvers -camping-ban-survey-says-police-dont-help.

Meinhardt, K. A., and C. A. Gehring. "Disrupting Mycorrhizal Mutualisms: A Potential Mechanism by Which Exotic Tamarisk Outcompetes Native Cottonwoods." *Ecological Applications* 22, no. 2 (2012): 532–49.

Merchant, C. "Shades of Darkness: Race and Environmental History." *Environmental History* 8, no. 3 (2003): 380–94.

Mills, J. E. *The Adventure Gap.* Seattle, WA: Mountaineers Books, 2014. Source of "cultural barrier . . ."

Muir, J. *A Thousand-Mile Walk to the Gulf.* Boston: Houghton, 1916. Source of "would easily pick . . ."

———. *My First Summer in the Sierra.* Boston: Houghton, 1917. Source of "dark-eyed . . ." and "strangely dirty . . ."

———. *Steep Trails.* Boston: Houghton, 1918. Source of "bathed in the bright river," "last of the town fog," "brave and manly . . . and crime," "intercourse with stupid town . . . ," and "doomed . . ."

Negro Motorist Green Book. New York: Green, 1949.

Online Etymology Dictionary. "Ecology." www.etymonline.com/index.php?term =ecology.

Pinchot, G. *The Training of a Forester*. Philadelphia: Lippincott, 1914. Source of "pines and hemlocks . . . general and unfailing."

Revised Municipal Code of the City and County of Denver, Colorado. Chapter 38: Offenses, Miscellaneous Provisions, Article IV: Offenses Against Public Order and Safety, July 21, 2015. municode.com/library/co/denver/codes/code_of_ ordinances?nodeId-TITIIREMUCO_CH38OFMIPR_ARTIVOFAGPUORSA.

Roden, J. S., and R. W. Pearcy. "Effect of Leaf Flutter on the Light Environment of Poplars." *Oecologia* 93 (1993): 201–7.

Royal Society for the Protection of Birds. "Giving Nature a Home." www.rspb.org.uk (accessed July 28, 2016).

Scott, M. L., G. T. Auble, and J. M. Friedman. "Flood Dependency of Cottonwood Establishment Along the Missouri River, Montana, USA." *Ecological Applications* 7, no. 2 (1997): 677–90.

Shakespeare, W. *As You Like It*. 1623. Available at http://www.gutenberg.org/ebooks/1121.

Strayed, C. *Wild*. New York: A. A. Knopf, 2012. Source of "myself a different story . . ."

The Nature Conservancy. "What's the Return on Nature?" www.nature.org/photos -and-video/photography/psas/natures-value-psa-pdf.pdf

U.S. Code, Title 16: Conservation, Chapter 23: National Wilderness Preservation System.

Vandersande, M. W., E. P. Glenn, and J. L. Walworth. "Tolerance of Five Riparian Plants from the Lower Colorado River to Salinity Drought and Inundation." *Journal of Arid Environments* 49, no. 1 (2001): 147–59.

Williams, T. T. *When Women Were Birds: Fifty-four Variations on Voice*. New York: Sarah Crichton Books, 2014. Source of "growing beyond . . . ," "things that happen . . . ," and "our own lips speaking."

Wohlforth, C. "Conservation and Eugenics." *Orion* 29, no. 4 (July 1, 2010): 22–28.

Callery Pear

Anderson, L. M., B. E. Mulligan, and L. S. Goodman. "Effects of Vegetation on Human Response to Sound." *Journal of Arboriculture* 10 (1984): 45–49.

Aronson, M. F. J., F. A. La Sorte, C. H. Nilon, M. Katti, M. A. Goddard, C. A. Lepczyk, P. S. Warren, et al. "A Global Analysis of the Impacts of Urbanization on Bird and Plant Diversity Reveals Key Anthropogenic Drivers." *Proceedings of the Royal Society of London B: Biological Sciences* 281, no. 1780 (2014), doi:10.1098 /rspb.2013.3330.

Bettencourt, L. M. A. "The Origins of Scaling in Cities." *Science* 340, no. 6139 (2013): 1438–41.

Borden, J. *I Totally Meant to Do That*. New York: Broadway Paperbacks, 2011.

Buckley, C. "Behind City's Painful Din, Culprits High and Low." *New York Times*, July 12, 2013. www.nytimes.com/2013/07/12/nyregion/behind-citys-painful-din -culprits-high-and-low.html.

Calfapietra, C., S. Fares, F. Manes, A. Morani, G. Sgrigna, and F. Loreto. "Role of Bio-genic Volatile Organic Compounds (BVOC) Emitted by Urban Trees on Ozone Concentration in Cities: A Review." *Environmental Pollution* 183 (2013): 71–80.

Campbell, L. K. "Constructing New York City's Urban Forest." In *Urban Forests, Trees, and Greenspace: A Political Ecology Perspective*, edited by L. A. Sandberg, A. Bardekjian, and S. Butt, 242–60. New York: Routledge, 2014.

Campbell, L. K., M. Monaco, N. Falxa-Raymond, J. Lu, A. Newman, R. A. Rae, and E. S. Svendsen. *Million TreesNYC: The Integration of Research and Practice.* New York: New York City Department of Parks and Recreation, 2014.

Cortright, J. *New York City's Green Dividend.* Chicago: CEOs for Cities, 2010.

Crisinel, A.-S., S. Cosser, S. King, R. Jones, J. Petrie, and C. Spence. "A Bittersweet Symphony: Systematically Modulating the Taste of Food by Changing the Sonic Properties of the Soundtrack Playing in the Background." *Food Quality and Preference* 24, no. 1 (2012): 201–4.

Culley, T. M., and N. A. Hardiman. "The Beginning of a New Invasive Plant: A History of the Ornamental Callery Pear in the United States." *BioScience* 57, no. 11 (2007): 956–64. Source of "marvel."

de Langre, E. "Effect of Wind on Plants." *Annual Review of Fluid Mechanics* 40 (2008): 141–68.

Dodman, D. "Blaming Cities for Climate Change? An Analysis of Urban Greenhouse Gas Emissions Inventories." *Environment and Urbanization* 21, no. 1 (2009): 185–201.

Engels, S., N.-L. Schneider, N. Lefeldt, C. M. Hein, M. Zapka, A. Michalik, D. Elbers, A. Kittel, P. J. Hore, and H. Mouritsen. "Anthropogenic Electromagnetic Noise Disrupts Magnetic Compass Orientation in a Migratory Bird." *Nature* 509, no. 7500 (2014): 353–56.

Environmental Defense Fund. "A Big Win for Healthy Air in New York City." *Solutions,* Winter 2014, page 13.

Farrant-Gonzalez, T. "A Bigger City Is Not Always Better." *Scientific American* 313 (2015): 100.

Gick, B., and D. Derrick. "Aero-tactile Integration in Speech Perception." *Nature* 462, no. 7272 (November 26, 2009), doi:10.1038/nature08572.

Girling, R. D., I. Lusebrink, E. Farthing, T. A. Newman, and G. M. Poppy. "Diesel Exhaust Rapidly Degrades Floral Odours Used by Honeybees." *Scientific Reports* 3 (2013), doi:10.1038/srep02779.

Hampton, K. N., L. S. Goulet, and G. Albanesius. "Change in the Social Life of Urban Public Spaces: The Rise of Mobile Phones and Women, and the Decline of Aloneness over 30 Years." *Urban Studies* 52, no. 8 (2015): 1489–1504.

Li, H., Y. Cong, J. Lin, and Y. Chang. "Enhanced Tolerance and Accumulation of Heavy Metal Ions by Engineered *Escherichia coli* Expressing *Pyrus calleryana* Phytochelatin Synthase." *Journal of Basic Microbiology* 55, no. 3 (2015): 398–405.

Lu, J. W. T., E. S. Svendsen, L. K. Campbell, J. Greenfeld, J. Braden, K. King, and N. Falxa-Raymond. "Biological, Social, and Urban Design Factors Affecting

Young Street Tree Mortality in New York City." *Cities and the Environment* 3, no. 1 (2010): 1–15.

Maddox, V., J. Byrd, and B. Serviss. "Identification and Control of Invasive Privets (*Ligustrum* spp.) in the Middle Southern United States." *Invasive Plant Science and Management* 3 (2010): 482–88.

Mao, Q., and D. R. Huff. "The Evolutionary Origin of *Poa annua* L." *Crop Science* 52 (2012): 1910–22.

Nemerov, H. "Learning the Trees." In *The Collected Poems of Howard Nemerov*. Chicago: The University of Chicago Press, 1977. Source of "comprehensive silence."

Newman, A. "In Leafy Profusion, Trees Spring Up in a Changing New York." *New York Times*, December 1, 2014. www.nytimes.com/2014/12/02/nyregion/in-leafy-blitz-trees-spring-up-in-a-changing-new-york.html.

New York City Comptroller. "ClaimStat: Protecting Citizens and Saving Taxpayer Dollars: FY 2014–2015 Update." comptroller.nyc.gov/reports/claimstat/#treeclaims.

New York City Department of Environmental Protection. "Heating Oil." www.nyc.gov/html/dep/html/air/buildings_heating_oil.shtml (accessed May 16, 2016).

———. "New York City's Wastewater." www.nyc.gov/html/dep/html/wastewater/index.shtml (accessed July 22, 2015).

New York State Penal Law. Part 3, Title N, Article 240: Offenses Against Public Order. ypdcrime.com/penal.law/article240.htm.

Niklas, K. J. "Effects of Vibration on Mechanical Properties and Biomass Allocation Pattern of *Capsella bursa-pastoris* (Cruciferae)." *Annals of Botany* 82, no. 2 (1998): 147–56.

North, A. C. "The Effect of Background Music on the Taste of Wine." *British Journal of Psychology* 103, no. 3 (2012): 293–301.

Nowak, D. J., R. E. Hoehn III, D. E. Crane, J. C. Stevens, and J. T. Walton. "Assessing Urban Forest Effects and Values: New York City's Urban Forest." Resource Bulletin NRS-9, U.S. Department of Agriculture, Forest Service, Northern Research Station, Newtown Square, PA, 2007.

Nowak, D. J., S. Hirabayashi, A. Bodine, and E. Greenfield. "Tree and Forest Effects on Air Quality and Human Health in the United States." *Environmental Pollution* 193 (2014): 119–29.

O'Connor, A. "After 200 Years, a Beaver Is Back in New York City." *New York Times*, February 23, 2007. www.nytimes.com/2007/02/23/nyregion/23beaver.html.

Peper, P. J., E. G. McPherson, J. R. Simpson, S. L. Gardner, K. E. Vargas, and Q. Xiao. *New York City, New York Municipal Forest Resource Analysis*. Davis, CA: Center for Urban Forest Research, USDA Forest Service, Pacific Southwest Research Station, 2007.

Rosenthal, J. K., R. Ceauderueff, and M. Carter. *Urban Heat Island Mitigation Can Improve New York City's Environment: Research on the Impacts of Mitigation Strategies on the Urban Environment*. New York: Sustainable South Bronx, 2008.

Roy, J. 2015. "What Happens When a Woman Walks Like a Man?" *New York*, January 8, 2015.

Rueb, E. S. "Come On In, Paddlers, the Water's Just Fine. Don't Mind the Sewage." *New York Times*, August 29, 2013. www.nytimes.com/2013/08/30/nyregion/in-water-they-wouldnt-dare-drink-paddlers-find-a-home.html.

Sanderson, E. W. *Mannahatta: A Natural History of New York City*. New York: Abrams, 2009.

Sarudy, B. W. *Gardens and Gardening in the Chesapeake, 1700–1805*. Baltimore, MD: Johns Hopkins University Press, 1998.

Schläpfer, M., L. M. A. Bettencourt, S. Grauwin, M. Raschke, R. Claxton, Z. Smoreda, G. B. West, and C. Ratti. "The Scaling of Human Interactions with City Size." *Journal of the Royal Society Interface* 11, no. 98 (2014), doi:10.1098/rsif.2013.0789.

Spence, C., and O. Deroy. "On Why Music Changes What (We Think) We Taste." *i-Perception* 4, no. 2 (2013): 137–40.

Tavares, R. M., A. Mendelsohn, Y. Grossman, C. H. Williams, M. Shapiro, Y. Trope, and D. Schiller. "A Map for Social Navigation in the Human Brain." *Neuron* 87, no. 1 (2015): 231–43.

Taylor, W. *Agreement for South China Explorations*. Washington, DC: Bureau of Plant Industries, U.S. Department of Agriculture, July 25, 1916.

West Side Rag. "Weekend History: Astonishing Photo Series of Broadway in 1920." November 30, 2014. www.westsiderag.com/2014/11/30/uws-history-astonishing -photo-series-of-broadway-in-the-1920s.

Wildlife Conservation Society. "Welikia Project." welikia.org (accessed July 24, 2015).

Woods, A. T., E. Poliakoff, D. M. Lloyd, J. Kuenzel, R. Hodson, H. Gonda, J. Batchelor, G. B. Dijksterhuis, and A. Thomas. "Effect of Background Noise on Food Perception." *Food Quality and Preference* 22, no. 1 (2011): 42–47.

Zhao, L., X. Lee, R. B. Smith, and K. Oleson. "Strong Contributions of Local Background Climate to Urban Heat Islands." *Nature* 511, no. 7508 (2014): 216–19.

Zouhar, K. "*Linaria* spp." In "Fire Effects Information System," produced by U.S. Department of Agriculture, Forest Service, Rocky Mountain Research Station, Fire Sciences Laboratory, 2003. www.fs.fed.us/database/feis/plants/forb/linspp/ all.html.

Olive

Besnard, G., B. Khadari, M. Navascués, M. Fernández-Mazuecos, A. El Bakkali, N. Arrigo, D. Baali-Cherif, et al. "The Complex History of the Olive Tree: From Late Quaternary Diversification of Mediterranean Lineages to Primary Domestication in the Northern Levant." *Proceedings of the Royal Society of London B: Biological Sciences* 280, no. 1756 (2013), doi:10.1098/rspb.2012.2833.

Cohen, S. E. *The Politics of Planting*. Chicago: University of Chicago Press, 1993.

deMenocal, P. B. "Climate Shocks." *Scientific American*, September 2014, pages 48–53.

Diez C. M., I. Trujillo, N. Martinez-Urdiroz, D. Barranco, L. Rallo, P. Marfil, and B. S. Gaut. "Olive Domestication and Diversification in the Mediterranean Basin." *New Phytologist* 206, no. 1 (2015), doi:10.1111/nph.13181.

Editors of the Encyclopædia Britannica. "Baal." *Encyclopædia Britannica Online*, last updated February 26, 2016. www.britannica.com/topic/Baal-ancient-deity.

Fernández, J. E., and F. Moreno. "Water Use by the Olive Tree." *Journal of Crop Production* 2, no. 2 (2000): 101–62.

Forward and Y. Schwartz. "Foreign Workers Are the New Kibbutzniks." *Haaretz*, September 27, 2014. www.haaretz.com/news/features/1.617887.

Friedman, T. L. "Mystery of the Missing Column." *New York Times*, October 23, 1984.

Griffith, M. P. "The Origins of an Important Cactus Crop, *Opuntia ficus-indica* (Cactaceae): New Molecular Evidence." *American Journal of Botany* 91 (2004): 1915–21.

Hass, A. "Israeli 'Watergate' Scandal: The Facts About Palestinian Water." *Haaretz*, February 16, 2014. www.haaretz.com/middle-east-news/1.574554.

Hasson, N. "Court Moves to Solve E. Jerusalem Water Crisis to Prevent 'Humanitarian Disaster.'" *Haaretz*, July 4, 2015. www.haaretz.com/israel-news/.premium -1.664337.

Hershkovitz, I., O. Marder, A. Ayalon, M. Bar-Matthews, G. Yasur, E. Boaretto, V. Caracuta, et al. "Levantine Cranium from Manot Cave (Israel) Foreshadows the First European Modern Humans." *Nature* 520, no. 7546 (2015): 216–19.

International Olive Oil Council. *World Olive Encyclopaedia*. Barcelona: Plaza & Janés Editores, 1996.

Josephus. *Jewish Antiquities, Volume VIII: Books 18–19*. Translated by L. H. Feldman. Loeb Classical Library 433. Cambridge, MA: Harvard University Press, 1965. Source of "construction of an aqueduct . . . ," "and tens of thousands of men . . . ," and "inflicted much harder blows . . ."

Kadman, N., O. Yiftachel, D. Reider, and O. Neiman. *Erased from Space and Consciousness: Israel and the Depopulated Palestinian Villages of 1948*. Bloomington: Indiana University Press, 2015.

Kaniewski, D., E. Van Campo, T. Boiy, J. F. Terral, B. Khadari, and G. Besnard. "Primary Domestication and Early Uses of the Emblematic Olive Tree: Palaeobotanical, Historical and Molecular Evidence from the Middle East." *Biological Reviews* 87, no. 4 (2012): 885–99.

Keren Kayemeth LeIsrael Jewish National Fund. "Sataf: Ancient Agriculture in Action." www.kkl.org.il/eng/tourism-and-recreation/forests-and-parks/sataf -site.aspx.

Khalidi, W. *All That Remains: The Palestinian Villages Occupied and Depopulated by Israel in 1948*. Washington, DC: Institute for Palestine Studies, 1992.

Langgut, D., I. Finkelstein, T. Litt, F. H. Neumann, and M. Stein. "Vegetation and Climate Changes During the Bronze and Iron Ages (~3600–600 BCE) in the Southern Levant Based on Palynological Records." *Radiocarbon* 57, no. 2 (2015): 217–35.

Langgut, D., F. H. Neumann, M. Stein, A. Wagner, E. J. Kagan, E. Boaretto, and I. Finkelstein. "Dead Sea Pollen Record and History of Human Activity in the Judean Highlands (Israel) from the Intermediate Bronze into the Iron Ages (~2500–500 BCE)." *Palynology* 38, no. 2 (2014): 280–302.

Lawler, A. "In Search of Green Arabia." *Science* 345, no. 6200 (2014): 994–97.

Litt, T., C. Ohlwein, F. H. Neumann, A. Hense, and M. Stein. "Holocene Climate Variability in the Levant from the Dead Sea Pollen Record." *Quaternary Science Reviews* 49 (2012): 95–105.

Lumaret, R., and N. Ouazzani. "Plant Genetics: Ancient Wild Olives in Mediterranean Forests." *Nature* 413, no. 6857 (2001): 700.

Luo, T., R. Young, and P. Reig. "Aqueduct Projected Water Stress Country Rankings." Washington, DC: World Resources Institute, 2015. www.wri.org/sites/default /files/aqueduct-water-stress-country-rankings-technical-note.pdf.

Neumann, F. H., E. J. Kagan, S. A. G. Leroy, and U. Baruch. "Vegetation History and Climate Fluctuations on a Transect Along the Dead Sea West Shore and Their Impact on Past Societies over the Last 3500 Years." *Journal of Arid Environments* 74 (2010): 756–64.

Perea, R., and A. Gutiérrez-Galán. "Introducing Cultivated Trees into the Wild: Wood Pigeons as Dispersers of Domestic Olive Seeds." *Acta Oecologica* 71 (2015): 73–79.

Pope, M. H. "Baal Worship." In *Encyclopaedia Judaica*, 2nd ed., vol. 3, edited by F. Skolnik and M. Berenbaum, pages 9–13. New York: Thomas Gale, 2007.

Prosser, M. C. "The Ugaritic Baal Myth, Tablet Four." Cuneiform Digital Library Initiative. cdli.ox.ac.uk/wiki/doku.php?id=the_ugaritic_baal_myth.

Ra'ad, B. *Hidden Histories: Palestine and the Eastern Mediterranean*. London: Pluto, 2010.

Snir, A., D. Nadel, and E. Weiss. "Plant-Food Preparation on Two Consecutive Floors at Upper Paleolithic Ohalo II, Israel." *Journal of Archaeological Science* 53 (2015): 61–71.

Stein, M., A. Torfstein, I. Gavrieli, and Y. Yechieli. "Abrupt Aridities and Salt Deposition in the Post-Glacial Dead Sea and Their North Atlantic Connection." *Quaternary Science Reviews* 29, no. 3 (2010): 567–75.

Terral, J., E. Badal, C. Heinz, P. Roiron, S. Thiebault, and I. Figueiral. "A Hydraulic Conductivity Model Points to Post-Neogene Survival of the Mediterranean Olive." *Ecology* 85, no. 11 (2004): 3158–65.

Tourist Israel. "Sataf." www.touristisrael.com/sataf/2503/ (accessed November 29, 2015).

Waldmann, N., A. Torfstein, and M. Stein. "Northward Intrusions of Low- and Mid-latitude Storms Across the Saharo-Arabian Belt During Past Interglacials." *Geology* 38, no. 6 (2010): 567–70.

Weiss, E. "'Beginnings of Fruit Growing in the Old World': Two Generations Later." *Israel Journal of Plant Sciences* 62 (2015): 75–85.

Zhang, C., J. Gomes-Laranjo, C. M. Correia, J. M. Moutinho-Pereira, B. M. Carvalho Goncalves, E. L. V. A. Bacelar, F. P. Peixoto, and V. Galhano. "Response, Tolerance and Adaptation to Abiotic Stress of Olive, Grapevine and Chestnut in the Mediterranean Region: Role of Abscisic Acid, Nitric Oxide and MicroRNAs." In *Plants and Environment*, edited by H. K. N. Vasanthaiah and D. Kambiranda, pages 179–206. Rijeka, Croatia: InTech, 2011.

Japanese White Pine

Auders, A. G., and D. P. Spicer. *Royal Horticultural Society Encyclopedia of Conifers: A Comprehensive Guide to Cultivars and Species*. Nicosia, Cyprus: Kingsblue, 2013.

Batten, B. L. "Climate Change in Japanese History and Prehistory: A Comparative Overview." Occasional Paper No. 2009-01, Edwin O. Reischauer Institute of Japanese Studies, Harvard University, 2009.

Chan, P. *Bonsai Masterclass*. Sterling: New York, 1988.

Donoghue, M. J., and S. A. Smith. "Patterns in the Assembly of Temperate Forests Around the Northern Hemisphere." *Philosophical Transactions of the Royal Society B: Biological Sciences* 359, no. 1450 (2004): 1633–44.

Fridley, J. D. "Of Asian Forests and European Fields: Eastern US Plant Invasions in a Global Floristic Context." *PLoS ONE* 3, no. 11 (2008): e3630.

Gorai, S. "Shugendo Lore." *Japanese Journal of Religious Studies* 16 (1989): 117–42.

National Bonsai & Penjing Museum. "Hiroshima Survivor." www.bonsai-nbf.org /hiroshima-survivor.

Nelson, J. "Gardens in Japan: A Stroll Through the Cultures and Cosmologies of Landscape Design." *Lotus Leaves, Society for Asian Art* 17, no. 2 (2015): 1–9.

Omura, H. "Trees, Forests and Religion in Japan." *Mountain Research and Development* 24, no. 2 (2004): 179–82.

Slawson, D. A. *Secret Teachings in the Art of Japanese Gardens: Design Principles, Aesthetic Values*. New York: Kodansh, 2013. Source of "if you have not received . . ."

Takei, J., and M. P. Keane. *Sakuteiki, Visions of the Japanese Garden: A Modern Translation of Japan's Gardening Classic*. Rutland, VT: Tuttle, 2008. Source of "wild nature" and "past master."

Voice of America. "Hiroshima Survivor Recalls Day Atomic Bomb Was Dropped." October 30, 2009. www.voanews.com/content/a-13-2005-08-05-voa38-67539217 /285768.html.

Yi, S., Y. Saito, Z. Chen, and D. Y. Yang. "Palynological Study on Vegetation and Climatic Change in the Subaqueous Changjiang (Yangtze River) Delta, China, During the Past About 1600 Years." *Geosciences Journal* 10, no. 1 (2006): 17–22.

Index